NEITHER
STAR WARS
NOR
SANCTUARY

NEITHER
STAR WARS
NOR
SANCTUARY

CONSTRAINING THE
MILITARY USES OF SPACE

MICHAEL E. O'HANLON

BROOKINGS INSTITUTION PRESS
WASHINGTON, D.C.

Copyright © 2004
THE BROOKINGS INSTITUTION
1775 Massachusetts Avenue, N.W., Washington, D.C. 20036
www.brookings.edu

Library of Congress Cataloging-in-Publication data
O'Hanlon, Michael E.
  Neither Star Wars nor sanctuary : constraining the military uses
of space / Michael E. O'Hanlon.
    p.   cm.
  Includes bibliographical references and index.
  ISBN 0-8157-6456-1 (cloth : alk. paper) —
  ISBN 0-8157-6457-x (pbk. : alk. paper)
  1. Space warfare. 2. Astronautics, Military—United States.
3. Ballistic missile defenses—United States. 4. United States—
Military policy. 5. World politics—21st century. I. Title.
UG1530.O33 2004   2004004467
358'.8—dc22

9 8 7 6 5 4 3 2 1

The paper used in this publication meets minimum requirements of the
American National Standard for Information Sciences—Permanence of
Paper for Printed Library Materials: ANSI Z39.48-1992.

Typeset in Sabon

Composition by R. Lynn Rivenbark
Macon, Georgia

Printed by R. R. Donnelley
Harrisonburg, Virginia

To Grace and Lily

and their mom

CONTENTS

Foreword   *ix*

Acknowledgments   *xiii*

1 Introduction   *1*

2 A Brief Primer on Space and Satellites   *29*

3 Current Threats and Technology Trends   *61*

4 A Future Taiwan Strait Conflict   *91*

5 Arms Control in Space   *105*

6 Preserving U.S. Dominance while Slowing the Weaponization of Space   *119*

Notes   *143*

Index   *165*

FOREWORD

Land, sea, and air—those are three environments where wars are waged. That is why the United States, like most countries, has three branches of armed forces: an army, a navy, and an air force. But with the onrush of technology, a new battlefield looms: outer space. So far, no nation stations weapons in space or deploys effective weapons on Earth with the intended purpose of threatening satellites. In the jargon that Mike O'Hanlon understands so well (and, mercifully, avoids in his own prolific writing), while space is not "weaponized" it *is* militarized. Today the U.S. Department of Defense uses space assets far more than ever before. Satellites, employed primarily in the cold war for strategic communications and for nuclear targeting and arms control monitoring, are now very much the instrument of the tactical warfighter. Weapons are guided to target by satellite, commanders plan their tactical operations using satellite links, real-time battlefield information is conveyed via satellite, enemy forces are located and attacked in large part on the basis of information gained and transmitted via U.S. space assets.

Other countries, including even the most technologically sophisticated of the United States' allies, cannot yet do most of these things. But they are gaining more and more expertise at monitoring Earth and communicating using satellites. New technologies, such as small satellites, are in the works as well, not only in Europe and Japan but also in China, India, Russia, and elsewhere. And ballistic missile defense programs, especially in the United States but also abroad, are gaining latent capabilities that could make them useful against low-Earth orbit satellites in particular—even as they are developed for much different reasons.

With the military uses of space expanding rapidly, there is a looming possibility that space will soon be truly weaponized, if only through the natural course of ongoing events. American military doctrine seems to take this trend for granted, even though no president has formally endorsed it as a desirable objective, nor has the U.S. government and strategic community carefully weighed the likely costs and benefits. An important part of what we do at Brookings is, precisely, assessing the costs and benefits of public policy choices facing the nation. It is Mike's premise that the weaponization of space would be a development of such consequence that the United States and other countries must make conscious decisions about whether to move in that direction, rather than simply drifting into what would be a revolutionary state of affairs.

Mike's own view is that space weaponry, under current and foreseeable circumstances, is neither necessary nor wise. However, he can imagine the situation changing—partly because military activities in space are hard to verify, making lasting bans on space weaponry impractical; partly because as other countries gain more military capability in space, their satellites may threaten American interests; and partly due to progress in ballistic missile defense technologies, with their inherent capabilities against low-Earth orbit satellites. Mike offers his own guidelines on how to reconcile sound judgments about the present with the uncertainties of the future, arguing for policies that deliberately attempt to slow movement toward the weaponization of space where possible, without foreclosing options for the United States in the future.

Brookings is grateful to the National Defense University and to the MacArthur Foundation for financial support of this project.

STROBE TALBOTT
President

Washington, D.C.
March 2004

ACKNOWLEDGMENTS

This book benefited enormously from the help of many individuals. Adriana Lins de Albuquerque and Aaron Moburg Jones provided excellent research assistance. General Mike Hamel and his colleagues at Vandenberg Air Force Base, as well as Dr. Don Daniel of the National Defense University, provided very important guidance as well (though, of course, they should not be held accountable for any opinions or mistakes in the manuscript, which are entirely the author's doing). Others who kindly assisted the author include Gordon Adams, Theresa Hitchens, Michael Krepon, Michael Levi, Jan Lodal, David Mosher, and a number of experts at Kirtland Air Force Base. Richard Garwin, Lt. Col. Peter Hays, James Clay Moltz, and James Steinberg reviewed the manuscript carefully and extremely helpfully. Finally, the author would like to acknowledge his deep intellectual debt in this subject to, among others, Robert Bell, Bruce Blair, Tom Christensen, Andrew Krepinevich, Andrew Marshall, John Pike, Jeffrey Richelson, Walt Slocombe, Tom Stefanick, Frank von Hippel, Barry Watts, and most of all Ashton Carter, Richard Garwin, and Paul Stares.

"Our nation may find it necessary to disrupt, degrade, deny or destroy enemy space capabilities in future conflicts. . . . U.S. Space Command currently does not have an operational anti-satellite weapon. . . . Research and development into anti-satellite technology is continuing."

U.S. SPACE COMMAND
November 26, 2001

"The Outer Space Treaty contains one over all rule: space shall be preserved for peaceful purposes for all countries."

GEORGE BUNN AND JOHN B. RHINELANDER
June 2002

CHAPTER ONE

INTRODUCTION

What future role will space play in warfare? And what should the United States do about it now? These questions have not been the focus of intensive and sustained political debate since the cold war days of the 1980s. In the meantime, technology has changed a great deal; geopolitics has changed even more. This book attempts to answer these broad questions for the context in which military space policy will be made in the early years of the twenty-first century.

Space is already militarized. Indeed, it has been militarized for more than four decades. But satellites played a rather benign role during much of the cold war, when they were most important for preserving strategic stability. Particularly since the cold war ended, however, space assets have been reestablished as competitive military instruments, especially by the United States. This trend has not extended to placing weapons in space or developing weapons for the purpose of threatening objects in space, but that clearly could change in the coming years. And weapons now being developed for other purposes, most notably missile defense, will make low-altitude satellites increasingly vulnerable even if no explicit steps are taken to achieve that end.

The Soviet Union and the United States employed satellites during most of the cold war. They did so largely for purposes of watching each other's nuclear tests, missile launches, and military force deployments. They also used space for communicating with their own global force deployments and operations, weather forecasting, mapping, measuring Earth's gravitational field (largely to improve the accuracy of ballistic missiles), and maintaining exact and uniform time standards for their deployed military forces. Many of these activities ultimately served the nonconfrontational and desirable purposes of maintaining strategic nuclear stability and promoting arms control. But their purposes were still basically quite military—contributing, for example, to the development of nuclear war plans—and hence were competitive as well. Indeed, from the launching of *Sputnik* in October 1957 until 1963, when a series of UN resolutions, implicitly at least, acknowledged and allowed the use of reconnaissance satellites, the Soviet Union struggled with the question of whether to tolerate U.S. satellites over its own territory. Both superpowers ultimately concluded that mutual toleration served their interests. The United States wanted means to tie together its global force deployments and to monitor capabilities in closed societies like the Soviet Union and the People's Republic of China (PRC). The Soviet Union saw its space program as a sign of national prestige and may have found reconnaissance satellites quite useful for watching events in places such as Cuba, China, and Europe.[1]

As time went on, both sides explicitly agreed not to interfere with the operations of each other's satellites in a number of arms control accords, including the 1972 ABM Treaty, the 1974 Threshold Test Ban Treaty, the 1976 Peaceful Nuclear Explosions Treaty, the 1979 SALT II Treaty, the 1990 multilateral CFE Treaty, and the 1991 START accords.[2] (They also signed the 1992 Open Skies Treaty, along with a couple dozen European countries, providing mechanisms for aerial monitoring under specific circumstances.)

Since the cold war, the United States has increasingly used satellite assets for tactical warfighting purposes in wars against Iraq, Serbia, and the Taliban in Afghanistan. Space systems, notably the global positioning system (GPS) satellite constellation, were used to help American soldiers navigate in the featureless desert starting most notably in the 1991 war against Iraq. GPS satellites are employed to synchronize operations in time as well, with remarkable accuracy. They are also increasingly used to pinpoint the locations of enemy targets and help guide precision-strike munitions, such as cruise missiles and the GPS-guided joint direct attack munition (JDAM), to those targets. Hundreds of JDAMs were used in the Kosovo war of 1999. More than 5,000 were employed in the Afghanistan war of 2001–02, striking as close as five meters from their aimpoints, and a comparable number were used in Operation Iraqi Freedom in 2003.[3] GPS devices are also integral to the "blue force" tracking systems that keep tabs on friendly units in a given region to reduce fratricide. Such systems still have only limited capabilities and use, and present challenges for filtering data so that users are not swamped by information they do not need, but they are quite useful nonetheless.[4]

Communications satellites are used for an increasing range of activities as well. While they still carry traditional voice messages, they also transmit real-time imagery taken by cameras and radar on platforms such as unmanned aerial vehicles

(UAVs) and reconnaissance aircraft to individuals far removed from the scene of battle, whether for purposes of data processing or for command and control.[5] They transmit detailed air war targeting plans to commanders and pilots.

As a result, the use of such satellites in war has skyrocketed. In Desert Storm, a total of sixteen military satellites and five commercial satellites provided coalition forces with a maximum possible transmission rate of 200 million bits per second (the equivalent of nearly 40,000 simultaneous telephone calls). Nearly twice as much capacity was available during the Kosovo war eight years later—much of it commercial, however, and hence unhardened against possible enemy action, such as electronic jamming, and unsecured. It was used for purposes that included teleconferencing among commanders.[6] Available capacity doubled again, to close to a billion bits per second, during the Afghan campaign of 2001–02. Again, much of the data flowed through commercial systems.[7] What that means is that, remarkably, a U.S. military operation of some 50,000 troops in 2001–02 used five times as much communications bandwidth as did a war with 500,000 troops a decade earlier—fifty times as much bandwidth per person, on average. In Operation Iraqi Freedom, the military used 2.4 gigabits per second.[8]

But the 2003 Iraq war was less notable for further increases in bandwidth, perhaps, than for several other aspects. More than fifty satellites were used in the war effort; commercial firms, including France's leading satellite services company, provided the majority of communications capacity and a fair amount of imagery as well. Satellite channels in the so-called EHF frequency band gave ships fifty times more bandwidth for secure data transmissions than in the past (128 kilobits per second). And the GPS permitted the United States to drop more than 6,000 satellite-guided JDAMs.[9]

Recognizing what satellites now offer the warfighter, the U.S. military is improving its means for utilizing their services. A space team was established and put on full-time duty in the Persian Gulf in late 2002 to plan operations against Iraq, for example. Among other things, its purpose was to help air planners understand when the greatest number of GPS satellites would be available to help guide bombs to target as accurately as possible.[10]

Space systems may soon be used to maintain a track on ballistic missiles, so that ground-based interceptors can be launched to shoot them down. Further in the future, space-based weapons may be used to destroy the ballistic missiles directly, though this is not necessarily a desirable goal for American policymakers anytime soon, as discussed below.

The increasing militarization of space is not exclusively a superpower story, however. The United States certainly dominates military space spending—accounting for more than 90 percent of the total, by some measures.[11] U.S. space funding over time is reported in table 1-1; the country's military space budget totals exceed $15 billion a year.[12] But other countries besides the United States and Russia have also increasingly sought military satellites, largely for reconnaissance and communications purposes so far, and will surely continue to pursue space capabilities of many types in the future. They may make use of civilian and commercial assets for military purposes as well. They are surely studying American capabilities to find, track, and quickly attack targets using space assets. Some are trying to emulate the United States; some are trying to find vulnerabilities in U.S. space systems so they can challenge them in any future wars. China may be the most notable example of a country that is doing both. Its progress to date is limited, as far as we can tell, and its progress in the coming years is likely to

Table 1-1. U.S. Government Space Funding

Billions of 2002 dollars

Fiscal Year	NASA	Department of Defense	Other	Total
1959	1.3	2.4	0.2	4.0
1960	2.3	2.8	0.2	5.3
1961	4.6	4.0	0.3	8.9
1962	8.8	6.3	1.0	16.1
1963	17.5	7.5	1.2	26.2
1964	23.9	7.6	1.0	32.5
1965	24.1	7.4	1.1	32.7
1966	23.4	7.8	1.0	32.2
1967	21.9	7.5	0.9	30.3
1968	19.4	8.4	0.7	28.6
1969	16.2	8.5	0.7	25.4
1970	14.3	6.8	0.6	21.7
1971	11.9	5.8	0.6	18.3
1972	11.2	5.1	0.5	16.9
1973	10.8	5.7	0.5	17.0
1974	9.2	5.9	0.5	15.7
1975	9.1	5.9	0.5	15.5
1976	11.5	6.9	0.6	19.0
1977	8.8	6.2	0.5	15.5
1978	8.9	6.7	0.6	16.2
1979	9.3	7.0	0.6	16.8
1980	10.0	8.2	0.5	18.6
1981	9.8	9.4	0.5	19.6

(continued)

Table 1-1. U.S. Government Space Funding (continued)

Billions of 2002 dollars

Fiscal Year	NASA	Department of Defense	Other	Total
1982	9.8	11.9	0.6	22.3
1983	10.5	15.0	0.5	26.1
1984	10.9	16.3	0.6	27.8
1985	10.6	19.6	0.9	31.2
1986	10.7	21.0	0.7	32.4
1987	14.3	23.7	0.7	38.6
1988	11.8	25.0	1.0	37.8
1989	13.8	24.5	0.8	39.1
1990	15.11	20.6	0.7	36.4
1991	16.6	18.0	1.0	35.6
1992	16.1	18.4	1.0	35.5
1993	15.6	16.8	0.9	33.2
1994	15.1	15.3	0.7	31.2
1995	14.3	12.1	0.9	27.3
1996	14.0	12.8	0.9	27.8
1997	13.6	12.8	0.9	27.3
1998	13.2	13.3	0.9	27.4
1999	13.2	14.0	1.0	28.2
2000	13.1	13.5	1.1	27.7
2001	13.6	14.7	1.1	29.4
2002	13.9	15.7	1.2	30.8
Total	568.1	505.1	32.9	1,106.2

Source: Tamar A. Mehuron, "2003 Space Almanac," *Air Force Magazine* (August 2003), p. 28.

be modest as well—but these prognostications may prove wrong, and in any case will not be applicable forever.

Although space is becoming increasingly militarized, it is not yet weaponized—at least as far as we know. That is, no country deploys destructive weapons in space, for use against space or Earth targets, and no country possesses ground-based weapons designed explicitly to damage objects in space. The challenges of weaponizing space should not be overlooked; in the words of one top Air Force specialist, space is a very challenging environment in which to work.[13] It is also a very different medium than the air, as Air Force Chief of Staff General John Jumper emphasized when he discarded the popular term "aerospace" and instead insisted that the Air Force must specialize in both air *and* space operations.[14] On the other hand, trends in technology and the gradual spread of space capabilities to many countries will surely threaten the status quo. Not only the United States but other major western powers, China, and smaller states as well, will have weaponization opportunities within reach.

But space is not a true sanctuary from weapons today. Virtually any country capable of putting a nuclear weapon into low-Earth orbit (LEO) already has a latent, if crude, antisatellite (ASAT) capability (though in many cases such weapons would have to be modified so that the warheads could be detonated by a timer or by remote control). Not only would such a weapon be likely to physically destroy any satellite within tens of kilometers of the point of detonation and to damage or destroy unhardened satellites within line of sight many hundreds of kilometers away (if not even further); it would also populate the Van Allen radiation belts with many more charged particles, which would destroy most low-Earth orbit satellites within about a month.[15]

Nor has space been treated as an inviolable sanctuary in the past. The nuclear superpowers made some progress toward developing antisatellite weapons in fits and starts from the 1950s through the 1980s. For example, the United States had something of an ASAT capability with its Nike Zeus and Thor nuclear-armed interceptor missiles in the 1960s and early 1970s, and with the Spartan program of the 1970s. The Soviets later developed and tested a nonnuclear "co-orbital" ASAT that needed to conduct a couple orbits to gradually approach its target (see table 1-2). Into the 1980s, the United States developed a nonnuclear "direct ascent" ASAT, launched by an F-15, that would reach its target much more promptly and then collide with it.[16] Soviet antiballistic missile (ABM) systems deployed around Moscow probably had ASAT capability as well; given the size of their warheads, they may have been able to damage satellites as distant as hundreds of kilometers from their detonation points.[17] Some of these capabilities may remain warehoused in some form. Still, the ASAT competition was held in check. Likewise, technological constraints made any deployment of space-based ballistic missile defenses impractical, even though the idea of such missile defenses was hotly debated.[18]

Decisions not to deploy ASATs or space-based missile defenses during the cold war did not, however, reflect any permanent commitment to keep space forever free from weaponry. Nor do existing arms control treaties ban such activities. Instead, they ban the deployment or use of nuclear weapons in outer space, prevent colonization of heavenly bodies for military purposes, and protect the rights of countries to use space to verify arms control accords and conduct peaceful activities.[19] In addition, in 2000 the United States and Russia agreed to notify

Table 1-2. Soviet Antisatellite Tests, 1968–82[a]

Test number and date	Target orbit			Intercept orbit				
	Target	Inclination (degrees)	Perigee; apogee (km)	Inter-ceptor	Inclination (degrees)	Perigee; apogee (km)	Attempted intercept altitude (km)	Probable outcome
Phase I								
1. Oct. 20, 1968	K248	62.25	475; 542	K249	62.23	502; 1,639	525	Failure
2. Nov. 1, 1968	K248	62.25	473; 543	K252	62.34	535?; 1,640?	535	Success
3. Oct. 23, 1970	K373	62.93	473; 543	K374	32.96	530; 1,053	530	Failure
4. Oct. 30, 1970	K373	62.92	466; 555	K375	62.86	565; 994	535	Success
5. Feb. 25, 1971	K394	65.84	572; 614	K397	65.76	575?; 1,000?	585	Success
6. Apr. 4, 1971	K400	65.82	982; 1,006	K404	65.74	802; 1,009	1,005	Success
7. Dec. 3, 1971	K459	65.83	222; 259	K462	65.88	231; 2,654	230	Success
Phase II								
8. Feb. 16, 1976	K803	65.85	547; 621	K804	65.86	561; 618	575	Failure
9. Apr. 13, 1976	K803	65.86	549; 621	K814	65.9?	556?; 615?	590	Success
10. Jul. 21, 1976	K839	65.88	983; 2,097	K843	n.a.	n.a.	1,630?	Failure[b]

11. Dec. 27, 1976	K880	65.85	559; 617	K886	65.85	532; 1,266	570	Failure[c]
12. May 23, 1977	K909	65.87	993; 2,104	K910	65.86	465?; 1,775?	1,710	Failure
13. Jun. 17, 1977	K909	65.87	991; 2,106	K918	65.9?	245?; 1,630?	1,575?	Success[d]
14. Oct. 26, 1977	K959	65.83	144; 834	K961	65.8?	125?; 302?	150	Success
15. Dec. 21, 1977	K967	65.83	963; 1,004	K970	65.85	949; 1,148	995	Failure[c]
16. May 19, 1978	K967	65.83	963; 1,004	K1009	65.87	965; 1,362	985	Failure[c]
17. Apr. 18, 1980	K1171	65.85	966; 1,010	K1174	65.83	362; 1,025	1000	Failure[c]
18. Feb. 2, 1981	K1241	65.82	975; 1,011	K1243	65.82	296; 1,015	1005	Failure[c]
19. Mar. 14, 1981	K1241	65.82	976; 1,011	K1258	65.83	301; 1,024	1005	Success
20. Jun. 18, 1982	K1375	65.84	979; 1,012	K1379	65.84	537; 1,019	1005	Failure[c]

Source: Paul B. Stares, *Space and National Security* (Brookings, 1987), p. 86.

? = Uncertain; n.a. = not available.

a. All missions were of two revolutions, except tests 8, 9, 12, and 13, which were one revolution.

b. Apparently failed to enter intercept orbit.

c. Reportedly used new optical sensor.

d. Conflicting data exist for intercept orbit.

each other in advance of most space launches and ballistic missile tests.[20] Most other matters are still up for grabs. And the concept of space as a sanctuary will be increasingly difficult to defend or justify as space systems are used more and more to assist in the delivery of lethal ordnance on target.[21]

Some scholars, such as Ambassador Jonathan Dean, do argue that the START I, Intermediate-Range Nuclear Forces (INF), and multilateral CFE treaties effectively ban the use of ASATs by one signatory against any and all others, given the protection they provide to satellite verification missions. But these treaties were signed before imaging satellites entered their own as targeting assets for tactical warfighting purposes, raising the legal and political question of whether protection originally provided to a satellite for one, generally nonprovocative and stabilizing mission can be extended to its use in a more adversarial fashion. Moreover, no one argues that these treaties ban the development, testing, production, or deployment of ASATs.[22]

In the late 1980s and 1990s, debates over military space policy became less visible than they had been during the Reagan era and a number of periods during the cold war. Détente, and then the end of the cold war, defused the immediate argument for such systems. Bill Clinton's election in 1992 reinforced these strategic developments, among other things leading to a shift in missile defense efforts from strategic to theater systems, for which weapons based in space did not figure prominently (though some theater missile defense [TMD] systems could have capabilities against low-Earth orbit satellites). Even when Clinton reemphasized national missile defense in mid-decade, his plan called for land-based interceptors. Sensor technology was to be based in space, but other capabilities were not. Clinton also curtailed the development of a kinetic energy, or "hit-to-kill," antisatellite system that he inherited from George H. W. Bush, as well as a microsatellite program known as Clementine II, despite the efforts of Senator Robert Smith of New Hampshire and other conservatives.[23]

But Clinton did not stop technology in its tracks. Two of the missile defense systems he promoted steadily, the midcourse national missile defense program and the airborne laser theater missile defense program, continue to this day and have latent capability as ASATs (see chapter 3). Moreover, he allowed the use of the mid-infrared advanced chemical laser (MIRACL) in a test against a target in space, confirming that the United States may have at least a rudimentary capability of using that ground-based high-energy laser in an ASAT mode.[24] (Meanwhile, some work continued more quietly and is ongoing under President Bush. For example, the Army has reportedly been working on laser dazzlers to blind surveillance satellites and jammers to disrupt communications and surveillance satellites.[25] It also, again, has a kinetic energy ASAT program, though funding has been near nil for several years and the Pentagon leadership has decided not to request funding for a flight test in 2004.[26])

The election of George W. Bush as president, and, even more important, his decision to select Donald Rumsfeld as secretary of defense, made it likely that such efforts would accelerate. Just before he became secretary, Rumsfeld chaired a commission on the military uses of space that warned of a possible future "space Pearl Harbor" for the United States unless it took a wide range of defensive and offensive steps to better protect its security interests in the heavens.[27] The worry was that countries such as China and Iran, among others, would gradually get their hands on technologies, such as high-energy lasers or homing microsatellites, that could threaten U.S. space assets. But the secretary's thinking is not strictly defensive. Rumsfeld's major strategic plan as secretary of defense states, "The mission of space control is to ensure the freedom of action in space for the United States and its allies and, when directed, to deny such freedom of action to adversaries."[28]

It is possible to exaggerate the change that occurred in U.S. policy when the Bush administration came into power. During the Clinton era, Air Force leaders increasingly discussed space as a military theater like any other. They envisioned the day when the Air Force would become an air and space force, or even a space and air force.[29] And Rumsfeld's language quoted above resembles official statements on Clinton administration space policy. Consider this excerpt from Space Command's 1998 Long-Range Plan in regard to one option it would eventually wish to develop more fully:

> *Negation* means applying military force to affect an adversary's space capability by targeting ground-support sites, ground-to-space links, or spacecraft. Negation will be executed when prevention fails. High-priority targets include an enemy's ability to hold US and allied space systems at risk. Negation will evolve from current concepts, which emphasize terrestrial attacks on an adversary's ground nodes, to a full range of flexible and discriminate techniques against the most appropriate node. Acting under clear lines of authority and rules of engagement, USCINCSPACE will take actions necessary to meet the National Command Authorities' objectives and defend our nation's vital space interests. Actions will range from temporarily disrupting or denying hostile space systems to degrading or destroying them. Our objectives must consider third-party use, plausi-

ble deniability and how actions will add to debris or otherwise affect the environment.[30]

But in broader context, Rumsfeld's approach indeed seems more assertive than policies under Clinton.[31] To quote Peter B. Teets, undersecretary of the Air Force for acquisition and head of the National Reconnaissance Organization, the nation must develop "ways to get a vehicle rapidly off the pad to any orbit on short notice. . . . It is easy to see how such a responsive capability could be useful for rapid constellation replenishment and sustainment, but I leave it to your imagination . . . to find other ways to employ such a capability to achieve desired warfighting effects."[32] Little imagination is needed if one reads official doctrine, given its emphasis on disrupting, degrading, and, if necessary, destroying enemy space assets in future combat.[33] This approach also seems to have emboldened a number of Air Force officers to make more public statements about the inevitability of weaponizing space.[34] As one example, Brigadier General Pete Worden argues that small launchers could be useful to the U.S. military for, among other reasons, their ability to launch weapons on short notice against targets in space.[35] Certain specific actions have already affected the policy landscape quite directly as well. For example, the Bush administration's decision to withdraw from the ABM Treaty, an action that was publicly announced in December 2001 and officially put into effect in June 2002, opened up the legal possibility of space-based ballistic missile defenses. Eliminating Space Command as one of the country's ten unified commands and placing space functions under Strategic Command may also reflect an inclination to think about space as another theater of combat, rather than as a special, and possibly safeguarded, domain.[36]

Not all policymakers agree with Rumsfeld. For example, Senate Majority Leader Tom Daschle made a strong statement against such activities in 2001 and suggested that most other Democrats also opposed putting weaponry in space.[37] This position is probably rooted, at least in part, in the philosophical view that space should remain a natural preserve of all mankind. As such, it should be devoted to nonmilitary—or at least nonthreatening and nonoffensive—purposes. Beyond that ideological point, opponents of weaponization also make a practical national interest argument: as the world's principal space power today, the United States stands to lose the most from the widespread weaponization of space, since that outcome could jeopardize the communications and reconnaissance systems on which it so disproportionately depends.[38]

Opponents of weaponizing space also point to the world's growing economic dependence on satellites, and the risk of damaging those capabilities should weaponry be based or used above the atmosphere. Worldwide, commercial revenues from space ventures exceeded government spending on space activities in the late 1990s, reaching and then exceeding $50 billion a year. The spread of fiber-optic cable has actually reduced the relative importance of satellites in global telephone services, and global economic conditions caused a downward revision in forecasts for space services.[39] But nonetheless the global satellite business now involves more than 1,000 companies in more than fifty countries.[40]

Non-American opponents of weaponizing space make many of the same arguments. They also worry about a unilateralist America pursuing its own military advantage at the expense of other countries, most of which do not favor putting weapons in space. This dispute has much of its origins and motivation in the history of the ballistic missile defense debate, as well as the

ASAT debate of the 1980s. But it has taken on a new tone in what many view as an era of American unipolarity or hegemony. In recent years, China and Russia have been consistently vocal in their opposition to the weaponization of space and their desire for a treaty banning the testing, deployment, and use of such capabilities.[41] So have a number of U.S. allies, including Canada, which in 1998 proposed that the United Nations convene a committee on outer space in its Conference on Disarmament in Geneva.[42] The UN General Assembly has continued to pass resolutions, for more than twenty straight years, opposing the weaponization of space. In December 2001 it called for negotiations on a treaty to prevent an arms race in outer space at the Geneva Conference. (The vote passed by 156 to 0; the United States, Micronesia, Israel, and Georgia abstained.)[43] In 2001 China presented an incomplete draft treaty banning the weaponization of space, and in 2002 China and Russia jointly presented another draft that included bans on weapons based in space and on any use of weapons against objects in space.[44]

For most defense planners today, by contrast, developing more military applications for outer space is an important imperative. Much thinking about the so-called revolution in military affairs and defense transformation emphasizes space capabilities. Ensuring American military dominance in the coming years—which proponents tend to see as critical for global stability as well—will require that the United States remain well ahead of its potential adversaries technologically. For some defense futurists, the key requirement will be to control space, denying its effective use to U.S. adversaries while preserving the unfettered operation of American satellites that help make up a "reconnaissance-strike complex." Others favor an even more ambitious approach. Given that fixed bases on land and large assets such as ships are increasingly vulnerable to precision-strike weaponry and other enemy capabilities—or to the political opposition of allies such as Turkey, Saudi Arabia, and France, who have sometimes opposed use of their territories or airspaces for military operations—they favor a greater U.S. reliance on long-range strike systems. These include platforms in space.[45]

Advocates of space weaponry also argue that, in effect, space is already weaponized, at least in subtle ways. As noted, most medium-range and long-range rockets capable of carrying nuclear weapons constitute latent ASATs. Likewise, rockets and space-launch vehicles could probably be used to launch small homing satellites equipped with explosives and capable of approaching and then destroying another satellite. Such capabilities may not even require testing, or at least testing easily detectable from Earth. Advocates of weaponization further note that the United States is willing to use weapons to deny other countries wartime use of the atmosphere, oceans, and land, raising the question of why space should be a sanctuary when these other media are not. As Barry Watts put it, "Satellites may have owners and operators, but, in contrast to sailors, they do not have mothers."[46]

Specific military scenarios can bring these more abstract arguments into clearer focus. Consider just one possibility. If in a future Taiwan Strait crisis China could locate and target American aircraft carriers using satellite technology, the case for somehow countering those satellites through direct offensive action would be powerful. If jamming or other means of temporary disruption could not be shown to provide reliable interruption of China's satellite activities, outright destruction would probably be seriously proposed—and would not immediately be unreasonable as an option. Indeed, China may be taking steps in the direction of using satellites for such targeting purposes, for example, with the recent launch of a 155-mile-range antiship cruise missile that may eventually be able to receive navigational updates by satellite communication link.[47] Moreover, despite rhetorical and diplomatic opposition to the weaponization of space, China's military planners have also reportedly given thought to how they might attack U.S. military space systems. That is quite a natural reaction for any defense planner who thinks his country may have to take on the United States someday. But it also underscores the strong pressures toward the weaponization of space, given current trends.[48]

Although technological progress, the absence of arms control regimes banning most military uses of space, and the growing use of space for tactical warfighting purposes suggest that space may ultimately be weaponized, the issue is not a simple yes or no proposition. The nature of the weapons that might ultimately reach space, or affect space assets, is important. So does the timing of weaponization, and the state of great power relations when it occurs. Even if weaponization is indeed inevitable, in other words, when and how it happens may matter a great deal. Accordingly, even if most weapons activities are not banned categorically by treaty, reciprocal restraint by the major powers, together with some limited and formal prohibitions on activities in space, may make sense.

This book is designed to move beyond the ideological debate of whether or not space should be preserved as a nonweaponized sanctuary, and instead to develop and analyze a number of specific proposals for future U.S. space policy.[49] The analysis is technical as well as strategic. It considers military, warfighting issues as well as arms control and political matters. Missile defense is not discussed in detail—numerous studies already exist on that subject—but its linkages to the space weapons debate are central and unavoidable, and thus frequently invoked in these pages.[50]

These questions need to be answered, in part, because there is at present no official U.S. position on most of them. Or, to put it differently, there are various competing positions. The military's publicly stated views are quite assertive, even if its actual programs for moving ahead with the weaponization of space are generally restrained for the moment. Moreover, most possible moves toward weaponization are unconstrained by any arms control accords. The Outer Space Treaty of 1967 only bans a small set of activities—notably, nuclear weapons in space, as well as hostile colonization of the moon and the planets.

Slowing the Weaponization of Space

This book's basic argument is that space should not yet be weaponized by the United States. For a combination of technological and strategic reasons, however, it may not prove practical to sustain this policy indefinitely. Thus the United States should also avoid most types of formal arms control categorically prohibiting the weaponization of space, even as it seeks to delay the arrival of that weaponization indefinitely.

Slowing and delaying the weaponization of space may strike many as an unsatisfying policy. It neither establishes a clear legal and political red line against such activity nor endorses it. Presumably, one might contend, putting weapons in space is either good or bad. If bad, should it not be precluded permanently; and if good on national security grounds, should it not be pursued without guilt or reservation?

In point of fact, space weapons are not inherently good or bad. Unlike biological weapons or many types of land mines, they are not by nature inhumane; yet, unlike the next type of fighter jet or munition, they are not just the natural progression of military modernization. Their political significance is much greater than that of most types of weaponry.

In addition, the United States enjoys a remarkably favorable military position in space today, without suffering much political and strategic fallout for making major use of the heavens for military purposes. It should preserve that situation as long as possible. And it has no need to rush to change current circumstances to maximize its own military capabilities. Some concepts, such as space-based ballistic missile defense, while holding a certain inherent appeal, would be needlessly provocative and exceptionally uneconomical to pursue at present. The satellites of other countries (and private companies) are not yet militarily significant enough to warrant development of destructive antisatellite weapons.

Extreme positions that would either hasten to weaponize space or permanently rule it out are not consistent with technological realities and U.S. security interests. The 2001 report of the Commission on Outer Space, which warned of a possible space "Pearl Harbor" and implied that the United States needed to take many steps—including offensive ones—to address such a purportedly imminent risk, was alarmist. It exaggerated the likely space capabilities of other countries. In fact, only certain classes of satellites are potentially vulnerable to enemy action in the coming years, and it may be some time before that potential vulnerability becomes real. Moreover, the United States can take passive and defensive measures to reduce the associated risks—and to know more clearly if and when it is being challenged in space.

To proceed on the basis of worst-case assumptions and hasten development of ASAT capabilities would be to ignore the serious political and strategic consequences of any U.S. rush to weaponize the heavens. American satellites, so dominant today, could be put at risk by the countervailing actions of other countries more quickly than they would be otherwise. Even more important, relations with Russia and China, which have improved in recent years but remain fragile, could suffer. Even if the United States someday does put weapons in space or develop weapons against objects in space, timing matters in international politics. Witness how the events of September 11, 2001, and the focused personal diplomacy between Presidents George W. Bush and Vladimir Putin preserved good relations between the United States and Russia even after the United States' withdrawal from the ABM Treaty in June 2002—an event that could have seriously damaged bilateral relations if it had occurred only a little earlier. Today, weaponizing space could reinforce the image of a unilateralist United States too quick to reach for the gun and disinclined to heed the counsel of others. Given that almost all countries routinely support an annual UN resolution calling for a treaty outlawing the weaponization of space, and that most currently find the United States too ready to flex its military muscle, any near-term decision to weaponize space would be very bad timing.

By the same token, the dismissive attitude toward any and all space weapons evidenced by large elements of the arms control community is too purist. Space, as noted, is increasingly used for warfighting purposes, so it cannot be viewed as a true sanctuary from military activity even today—and it is not clear that space should be seen as a less appropriate place to fight than Earth. If satellites increasingly become tools of the tactical warfighter rather than linchpins of strategic stability between nuclear-armed powers, it is not clear that they should merit complete protection from attack even as they are used to help kill targets on the ground. Leaving aside philosophical arguments, there are also practical military rationales for keeping a future ASAT option. While the United States might like to preserve its current dominance in space for intelligence, communications, and tactical warfighting purposes, it will not enjoy that luxury forever. Passive steps to defend itself, such as satellite hardening, may not suffice to protect its interests—even in the hypothetical case of an ASAT treaty banning the development, deployment, and use of antisatellite weapons. Too many non-ASAT technologies have potential applications as ASATs, especially in a world of increasingly capable missile defenses and a growing number of satellites and microsatellites. And other countries may learn to use satellites for tactical warfighting, including against the United States.

Some would argue that missile defenses themselves are unwarranted. But in the international enviroment, the demise of the ABM Treaty is quite certainly now permanent, with no prospect of a replacement accord prohibiting such technologies. Even defenses designed against shorter range missiles (often known as theater missile defenses), which have not been controversial in the United States, have certain antisatellite capabilities. Moreover, a number of these technologies are close to realization in the United States. Regardless of whether one thinks it would be desirable, it is simply not feasible to put the horse back in the barn.

That said, the United States should pursue some types of binding arms controls on military space activities and, even more important, show unilateral restraint on its space activities in a number of ways. It should agree to a ban on any tests in space that would create debris (notably, tests of antisatellite weapons that use explosives or collisions to destroy targets). It should publicly declare that it will forgo space tests of any antisatellite system for the foreseeable future. And it should also seriously consider revising its military space doctrine to declare that it will not even develop dedicated ASAT technologies in the coming years. This policy will probably prove temporary, but because the coming years will be critical for the further maturation and improvement of great power relations (especially with Russia and China), improving the prospects for strategic stability in that period is important. If and when the United States needs to change its policy in the future, the danger of strategic fallout may be reduced.

The approach recommended here differs greatly from the early rhetorical position of the incoming Bush administration. It differs less from the de facto approach of the Bush administration since September 11, 2001, when the United States chose to emphasize the need for great power security cooperation against terrorism and to seek accommodation or delay on issues that could impede that priority effort. But even today, the Bush administration retains the aspiration for space-based missile defense and funds programs to that effect; it retains a space policy doctrine emphasizing the possibility of destroying the satellites of potential adversaries; it refuses to negotiate even very limited accords on uses of space that might, for example, prevent the production of more debris in low-Earth orbit; and it establishes no policy roadblocks to the rapid weaponization of space, should it choose to move in that direction in any new budget plan. It should make its de facto restraint more formal in certain areas and reinforce it in several others.

Translating these broad strategic observations and premises into policy terms leads to the set of core recommendations described below. They begin with more straightforward, passive, and nonthreatening actions that the United States could take to ensure its reliable use of military space assets in the future. But they also include options for the weaponization of space, particularly in regard to antisatellite capabilities of various types, should future circumstances so warrant.

Passive and Defensive Measures

—Beyond improving its ground-based space surveillance capabilities as currently planned, the United States could place surveillance assets on individual satellites to identify and report any attacks on them. Most military satellites lack such capabilities today.

—Although details are classified, the United States appears not to have sufficiently hardened its military satellites. It needs to do so not only against natural radiation and nuclear effects, but also against certain other threats; laser attacks against low-Earth orbit satellites are among the most worrisome. Hardening of new military satellites is generally feasible and practical, albeit not cost free.

—A related measure could be to subsidize hardening of commercial satellites (particularly communications systems) on which the United States increasingly depends. Jamming and nuclear-induced atmospheric disturbances are among the more serious threats to guard against. But these steps may not be practical given classification concerns, commercial satellites' emphasis on high-data-rate transmissions, and other factors. Thus the United States needs to continue to emphasize laser satellite communications systems and reduce its dependency on commercial satellites in warfighting environments. In the interim, the U.S. armed forces also need to be able to streamline their insatiable demands for data and bandwidth, because such wide bandwidths may not be reliably available in future conflicts. Specifically, the military needs to prioritize its data needs and develop mechanisms for ensuring that the most important information can continue to flow in combat even if 25 percent,

50 percent, or 75 percent of total desired bandwidth proves at least temporarily unavailable due to enemy action.

—The United States needs to be able to recover if major satellite capabilities, such as low-Earth orbit imaging assets, are damaged or destroyed. Many in the military community favor development of rapid reconstitution capabilities—extra satellites in warehouses, coupled with rockets ready to launch within weeks or even days of a decision to do so. But if satellites on orbit proved vulnerable, a second batch of satellites might be, too. Thus air-breathing capabilities, such as the P-3, EC-135, JSTARS, and various UAV systems need to be retained. The GPS constellation may be sufficiently robust and distributed that most of its satellites will survive any plausible attack, but the signals of current-generation GPS satellites are relatively easy to jam or otherwise disrupt. That may suggest the need for airborne targeting capabilities as a backup to GPS; more likely, it suggests that the United States needs to modernize its GPS system by putting into orbit an improved generation of satellites without further delay.

—The United States should research active defenses for satellites. These would not necessarily have to be general-purpose ASAT weapons; they could instead be short-range self-defense weapons placed on the satellite to be defended and designed to strike only nearby objects. Their kill mechanisms could, for example, be high-powered microwaves or lasers of modest total power.

Antisatellite Technologies

—Partly because the future survivability of its own satellites cannot now be assumed, and partly because the future survivability of adversary satellites may not be tolerable under certain circumstances, the United States should not rule out the possibility of developing ASAT capabilities of its own. It should not hasten to develop, test, or deploy advanced systems for this purpose. Yet nor should it preclude the possibility, either by treaty or by excessive constraints on its basic research and development activities.

—In fact, the United States will soon possess latent ASAT capability. These systems will not be found only in the form of nuclear-tipped intercontinental ballistic missiles (ICBMs) or submarine-launched ballistic missiles (SLBMs), which could be programmed to detonate at a certain time near a certain point in space; nor solely in the form of the MIRACL laser already operational in New Mexico. They will increasingly be found in ballistic missile defense programs as well. In particular, the midcourse intercept system soon to be deployed in Alaska and California surely has at least latent capability against low-Earth orbit satellites, even if it might require software upgrades to accept targeting data from different sensors than would likely be used for missile defense. And the airborne laser (ABL) will soon have similar capabilities. Again, the ABL would need help from external sensors to find and track a satellite, and quite likely would require software upgrades to be able to accept the data from those sensors. These types of software modifications, and actual testing of these weapons in an ASAT mode, should be avoided indefinitely. But their intrinsic ASAT capabilities against most types of lower-Earth orbit satellites appear rather significant.

—The United States should not build dedicated ASATs soon. It has enough advantages in any ASAT competition in the form of its ABL and midcourse ballistic missile defense programs that it need not be first in each and every technology category, including microsat ASATs. The downside of developing microsat ASATs first is that doing so would harm great power strategic relations and help accelerate an ASAT arms race that the United States does not stand to benefit from in the foreseeable future. Similarly, a kinetic energy ASAT is unnecessary, and would be undesirable vis-à-vis other possible ASAT technologies in any event, given the debris it would cause in space.

Missile Defenses and Space-to-Earth Weapons

—The United States should not hasten the development or deployment of space-based missile defenses, which would have inherent ASAT capabilities. They are not needed for missile defense against extremist states with modest arsenals—the only real rationale for ballistic missile defense systems in the foreseeable future—given the variety of ground-based options soon to be available. For purposes of missile defense they would have to be deployed in such numbers (given absentee ratios, due to the movement of satellites above Earth) that they could pose a very serious threat to many satellites simultaneously, as well as being extraordinarily expensive.

—Space-to-Earth weapons are not a promising concept for the foreseeable future. In addition to being politically very provocative, they offer few benefits to a global military power already capable of rapid intercontinental strike. The technologies within reach, such as tungsten rods or a common aero vehicle that could function first as a reentry vehicle and then as a guided aerodynamic device, do not warrant advanced development and deployment. They are either too limited in capabilities, too expensive, or too uninteresting in terms of their limited attributes relative to ground-based systems. Further conceptual exploration and basic research may be warranted; nothing more than that is even desirable in the coming years—and hence budgets need not be substantially increased.

CHAPTER TWO

A BRIEF PRIMER ON SPACE AND SATELLITES

Space is a remarkable and unique environment. It is also now a fairly heavily populated one. Near-Earth space is home to a wide range of military and civilian satellites, not to mention vast amounts of debris that can interfere with satellite operations. Assets in space also require assets on the ground, and links with the ground, to provide services to military users of satellites. This chapter explains some of the basic geometry and physics of space, rocketry, and satellite operations and surveys the existing capabilities in space of a number of nations and firms, as well as capabilities they intend to develop and orbit in the near future. It also notes trends that will tend to reduce the U.S. military dominance of space over time. The United States will not lose its large technological lead, but it is likely to lose the near-monopoly it has enjoyed in this sphere of military competition in recent years.

SPACE AND SATELLITES

Satellite Orbits

Most satellites move around Earth at distances ranging from 200 kilometers to about 36,000 kilometers. This region is divided into three main bands (see figure 2-1). Low-Earth orbit extends out to about 5,000 kilometers.[1] Geosynchronous orbit (GEO) is the outer band for most satellites. It is 35,888 kilometers or 22,300 miles above the equator of Earth. At that altitude, a satellite's revolution around the Earth takes exactly twenty-four hours, meaning that it remains over the same spot on Earth's equator continuously. Medium-Earth orbit (MEO) is essentially everything in between LEO and GEO. MEOs are concentrated between 10,000 and 20,000 kilometers above the surface of Earth.

The range of LEO orbits begins just above Earth's atmosphere, which is generally considered to end at an altitude of about 100 kilometers—though a considerable number of air molecules are found even higher, meaning that there is atmospheric drag on especially low LEO satellites. (Aircraft do not fly above 40–45 kilometers, so the region from 50 to 150 kilometers is "uninhabited.")[2] The altitude of LEO orbits is less than the radius of Earth (which is about 6,400 kilometers, or almost 4,000 miles). In other words, if one viewed low-altitude satellites from some distance, they would appear quite close to Earth (see figure 2-1). The dimensions of geosynchronous orbits are large relative to the size of Earth (though they are still small relative to the distance between Earth and the moon, about 380,000 kilometers). In other words, by the standards of the solar system, all artificial satellites are very close to Earth, meaning that any one can "see" only a small part of Earth's surface at a time (see table 2-1). They are also all clearly within the range and influence of Earth's gravitational field; they have

Figure 2-1. Major Orbit Types

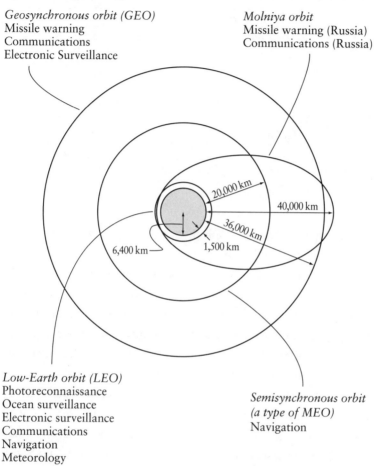

Geosynchronous orbit (GEO)
Missile warning
Communications
Electronic Surveillance

Molniya orbit
Missile warning (Russia)
Communications (Russia)

Low-Earth orbit (LEO)
Photoreconnaissance
Ocean surveillance
Electronic surveillance
Communications
Navigation
Meteorology

Semisynchronous orbit
(a type of MEO)
Navigation

Source: Ashton Carter, "Satellites and Anti-Satellites: The Limits of the Possible," *International Security*, vol. 10, no. 4 (Spring 1986), p. 49.

not been imparted with enough speed to escape it, as would be required for a space probe.[3] Earth's gravitational field, together with the velocity (speed and direction of movement) of a satellite, establish the parameters for that satellite's orbit. Once these

Table 2-1. Single Satellite Coverage Requirements in Polar Orbit[a]

Number of satellites	Altitude (miles)
60	300
50	400
30	600
25	800
20	1,000
12	2,000
3 (GEO)	23,000

Source: Paul B. Stares, *Space and National Security* (Brookings, 1987), p. 40.

a. Table shows the number of imaging satellites required to keep all of Earth in constant view.

physical parameters are specified, the orbit is determined and trajectories are predictable.

Close-in satellite orbits take as little as ninety minutes. As noted, geosynchronous orbits take exactly twenty-four hours. Satellites in close-in circular orbits move at nearly eight kilometers per second; those in geosynchronous orbit move at about three kilometers per second. Those following intermediate orbits have intermediate speeds and periods of revolution about Earth.

Satellite orbits are generally circular, though a number are elliptical, and some are highly elliptical—passing far closer to Earth in one part of their orbit than in another. Satellites may move in polar orbits, passing directly over the North and South Poles once in every revolution around Earth. Alternatively, they may orbit continuously over the equator, as do GEO satellites, or may move along an inclined path falling somewhere between polar and equatorial orientations.

Getting satellites into orbit is, of course, a very challenging enterprise. They must be accelerated to very high speeds and properly oriented in the desired orbital trajectories. Modifying a satellite's motion is very difficult once the rocket that puts it into space has stopped burning; generally, the satellite's own boosters are only capable of fine-tuning a trajectory, not changing it fundamentally. Even though satellites in GEO end up moving much more slowly than satellites in LEO, they must be accelerated to greater initial speeds. That is because they lose a great deal of speed fighting Earth's gravity as they move from close-in altitudes to roughly 36,000 kilometers above the planet's surface. If they are to reach the final GEO destination with enough speed to stay there in orbit, they require an initial speed near Earth's surface of about 10.5 kilometers per second, nearly enough to escape Earth's gravitational pull and its orbit altogether. By contrast, a rocket putting a satellite into LEO typically only reaches its final speed once it is nearing the desired altitude, so it would not exceed eight kilometers per second at any time.

When rocket propulsion is involved, the difference between 8 kilometers per second and 10.5 kilometers per second is much more than those numbers would suggest. Gaining the final 2.5 kilometers per second of speed requires enormous effort and fuel. A three-stage rocket that could carry a payload of fifteen tons into LEO, for example, could only transport three tons into GEO.[4] For that reason it typically costs two to three times as much per pound of payload to put a satellite into GEO as into LEO.[5] And even getting to LEO is difficult. For example, putting a payload into low-Earth orbit typically requires a rocket weighing 50 to 100 times as much as the payload.[6] Consequently, even low-Earth orbit launch is stubbornly expensive, despite longstanding efforts to reduce launch costs; putting a satellite into LEO typically costs from $3,000 to $6,000 per pound (though some Ukrainian and Chinese launch services charge less than $2,000).[7]

Most satellites weigh from 2,000 pounds to 10,000 pounds, roughly speaking, implying launch costs of about $10 million for smaller satellites in LEO to $100 million for larger satellites in GEO. There are exceptions, however, including the large imaging satellites known as Lacrosse and KH-11, each of which is believed to weigh about 30,000 pounds and to require a rocket with the capabilities of the enormous Titan IV to launch it (at a cost of $400 million). In addition, most satellites have dimensions ranging from 20 feet to 200 feet and power sources capable of generating 1,000 to 5,000 watts—though again, imaging satellites would be expected to exceed these bounds.[8]

A final note on this brief primer on space and satellites concerns the so-called Van Allen radiation belts, discovered by a scientist of that name in the late 1950s. Satellites operating in these regions require extra shielding to protect themselves from the electrons and protons that tend to be trapped in these zones over extended periods by Earth's magnetic fields. There are two belts, the inner one consisting largely of protons and most intense at about 3,500 kilometers above Earth, the outer consisting largely of electrons and peaking in intensity at about 16,000 kilometers. The inner belt begins somewhere between 400 kilometers and 1,200 kilometers, depending on latitude (and is most prevalent from 45 degrees north to 45 degrees south), and extends out to about 10,000 kilometers. The outer belt begins at roughly the latter distance from Earth and extends well beyond GEO, going as high as 80,000 kilometers, depending on the recent activity of the sun. Although the two belts can overlap, they are both weak in the region where they would do so. As a practical matter, therefore, satellites in the lowest LEOs as well as MEOs of roughly 10,000 kilometer altitude need not be shielded particularly strongly against the Van Allen belts.[9]

Current Military and Commercial Satellites

There are currently 8,000 to 9,000 objects in space that are large and visible enough to be tracked by U.S. monitoring equipment. Given the state of technology at present, that implies a diameter of at least ten centimeters (about four inches). Less than 1,000 of these objects are working satellites; the rest are old satellites or large pieces of debris from rockets (see table 2-2).[10]

In recent years, about a third of all launches have been from Russia and other former Soviet republics, just over a third from the United States, and just under a third from the rest of the world in aggregate. Commercial and military launches were placing about 150 satellites in orbit annually at the end of the 1990s, though only about sixty to sixty-five a year in 2001 through 2003. Projections posit a modest increase in the global market in coming years—in 2003 one source estimated that about 100 satellites would be orbited in 2004, 110 in 2005, and nearly 130 by 2007. The associated growth in the expected value of launched units was from $10.2 billion in 2004 to $11.8 billion in 2007 (in constant 2003 dollars).[11]

The recent downturn in the global market began before September 11, 2001. It was a reflection of cyclical economic factors, the continued high cost of launch, and the growing use of fiber-optic landline communications, producing a lack of growth in the market for satellite communications capabilities. Although considerable recovery is expected, as noted, the market is unlikely to take off anytime soon. Buttressing this conclusion, technological progress in launch vehicle technology is quite modest at present, and hopes for drastically lower launch prices remain stymied by the lack of radical technological breakthroughs in materials, fuels, or reusable rocket technologies (see chapter 3).[12]

Table 2-2. Objects in Earth Orbit, as of May 31, 2003

Country or organization[a]	Number of satellites[b]	Number of space probes	Pieces of debris	Total
Commonwealth of Independent States	1,338	35	2,550	3,923
United States	889	49	2,842	3,780
European Space Agency	33	2	305	340
People's Republic of China	36	0	282	318
India	25	0	124	149
Japan	80	6	56	142
International Telecom Satellite Organization	60	0	0	60
Globalstar	52	0	0	52
France	33	0	15	48
Orbcomm	35	0	0	35
European Telecom Satellite Organization	24	0	0	24
Germany	19	2	1	22
United Kingdom	21	0	1	22
Canada	20	0	1	21
Italy	11	0	3	14
Luxembourg	13	0	0	13
Australia	8	0	2	10
Brazil	10	0	0	10
Sweden	10	0	0	10
Indonesia	9	0	0	9
International Maritime Satellite Organization	9	0	0	9
NATO	8	0	0	8
Arab Satellite Communication Organization	7	0	0	7
Argentina	7	0	0	7
Sea Launch	1	0	6	7
South Korea	7	0	0	7

(continued)

Table 2-2. Objects in Earth Orbit, as of May 31, 2003 (continued)

Country or organization[a]	Number of satellites[b]	Number of space probes	Pieces of debris	Total
Mexico	6	0	0	6
Spain	6	0	0	6
Netherlands	5	0	0	5
Asia Satellite Corporation	4	0	0	4
Czech Republic	4	0	0	4
International Space Station	1	3	0	4
Israel	4	0	0	4
Thailand	4	0	0	4
Turkey	4	0	0	4
Malaysia	3	0	0	3
Norway	3	0	0	3
Saudi Arabia	3	0	0	3
Egypt	2	0	0	2
France and Germany	2	0	0	2
Philippines	2	0	0	2
Algeria	1	0	0	1
Chile	1	0	0	1
China and Brazil	1	0	0	1
Denmark	1	0	0	1
EUME	1	0	0	1
Greece	1	0	0	1
NICO	1	0	0	1
Pakistan	1	0	0	1
Portugal	1	0	0	1
Taiwan	1	0	0	1
Saudi Arabia and France	1	0	0	1
Singapore and Taiwan	1	0	0	1
United Arab Emirates	1	0	0	1
Total	2,831	97	6,188	9,116

Source: Tamar A. Mehuron, "2003 Space Almanac," *Air Force Magazine* (August 2003), p. 24.

a. The Commonwealth of Independent States includes Russia and the former Soviet Union. EUME is the European meteorological satellite.

b. Not all satellites are still functional.

The vast majority of most countries' current satellites are in LEO or GEO. In fact, excluding Russian satellites (with their particular history and their particular circumstances, servicing a large northern country), each of those zones accounts for about 45 percent of the 600 satellites in active use today.[13] Another 5 percent are in MEO; most of the remainder are located in highly elliptical orbits. Of the total of 600 non-Russian satellites, nearly 350 are used for general communications, 140 for military communications and imaging, 60 for navigation, and 50 for scientific or other commercial purposes.

In many cases, the dividing line between military and civilian satellites is blurred. The United States uses GPS satellites for military and civilian purposes. It buys time on commercial satellites for military communications, which now constitute well over half of all total satellite communication capacity.[14] Since 1994, when President Bill Clinton issued his presidential decision directive (PDD) 23, private U.S. companies have been allowed to operate imaging systems provided that the government can maintain "shutter control" in times of national crisis.[15] (PDD-23 was superceded in 2003 by the U.S. Commercial Remote Sensing Policy, which stated that the government would go even further and use commercial assets whenever possible, even for national security needs.) The U.S. military and intelligence services often purchase imagery from private firms, especially when relatively modest-resolution images (with correspondingly larger fields of view) are adequate (see figure 2-2). For example, during the war in Afghanistan in 2001, the National Imagery and Mapping Agency (NIMA) signed an agreement with the Space Imaging Corporation, reportedly for $1.9 million a month, obtaining exclusive access to its imagery through early January 2002 (perhaps in part to deny these services to others, including the United States' enemies).[16] And some satellites provide weather

Figure 2-2. Trends in Commercial Earth Imaging Satellites

Resolution (meters)

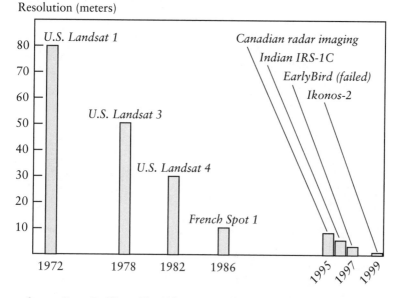

Source: Barry D. Watts, *The Military Uses of Space: A Diagnostic Assessment* (Washington: Center for Strategic and Budgetary Assessments, 2001), p. 66.

data to the military as well as to other government agencies. Indeed, the civilian satellite systems made use of by the U.S. armed forces form a lengthy list (see table 2-3).

Other countries do likewise. China deploys communications and reconnaissance satellites ostensibly under civilian pretexts, but there can be little doubt that its military has use of them and the data they create. Many countries purchase satellite imagery from commercial firms for their own national security purposes, or use commercial satellite communications links for a fee. Because of this blurring between military and commercial uses, the summary of global satellite capabilities that follows discusses all assets of a given country (or association of countries).

Table 2-3. Major Civilian Satellites in U.S. Military Use, as of 2003

Satellite	Mission/capabilities	First launch	Number in con- stellation	Orbit altitude (miles)	Weight (lbs)
Advanced Communications Technology Satellite (ACTS)	Demonstration satellite for new types of K- and Ka-band communications technologies	1993	1	22,300	3,250
Geostationary Operational Environmental Satellite (GOES)	Weather data collection for forecasting	1975	2	22,300	4,600
Globalstar	Mobile communications with security controls	1998	48	878	990
Ikonos	One-meter resolution Earth imaging	1999	1	423	1,600
Inmarsat	Peacetime mobile communications services	1990	9	22,300	4,545
Intelsat	Routine communications and distribution of Armed Forces Radio and TV Services network	1965	20	22,300	7,480
Iridium	Voice and data transmission for handheld mobile communications	1997	66	485	1,516

Landsat	Imagery	1972	1	438	4,800
Loral Orion	Rooftop-to-rooftop communications for U.S. Army	1994	3	22,300	8,360
NOAA 15 and NOAA 16 (NOAA or TIROS)	Weather worldwide	1978	2	517	4,900
Orbcomm	Mobile communications in Joint Interoperability Warfighter Program	1995	35	500–1,200	90
Pan Am Sat	Routine communications	1983	21	22,300	6,760
Quickbird 2	High-resolution imagery for mapping, military surveillance, weather research, and other uses	2001	1	279	2,088
Satellite Pour l'Observation de la Terre (SPOT)	Terrain imaging used for mission planning systems, terrain analysis, and mapping	1986	3	509	5,940
Tracking and Data Relay Satellite System (TDRSS)	Global network facilitating communication between LEO satellites and control stations without an elaborate ground network	1983	6	22,300	5,000

Source: Mehuron, "2003 Space Almanac," pp. 43–44.

Before proceeding to this discussion, a brief word is in order about the vast majority of man-made material in space. In short, it is junk, and dangerous junk at that. There are probably 100,000 pieces of debris larger than a marble in orbit; those at altitudes above 1,000 kilometers will remain in orbit for centuries, and those above 1,500 kilometers for millennia.[17] Perhaps 300,000 small objects, such as chips of metal or even specks of paint, are too small to be tracked—yet at least four millimeters in size, large enough to do potential harm to any object they would strike, given the enormous speeds of collision implied by orbiting objects. In 1983, for example, a paint speck only 0.2 millimeters in diameter made a 4 millimeter dent in the *Challenger* space shuttle's windshield.[18] Only two other collisions between debris and operational satellites were known to have occurred through 2001, but with debris in low orbital zones growing at the rate of about 5 percent annually, more can certainly be expected.[19] Indeed, a small satellite at an altitude of 800 kilometers now has about a 1 percent chance annually of failure due to collision with debris. And below 2,000 kilometers, there is now a total of 3 million kilograms of debris of various kinds (in contrast to about 200 kilograms of meteoroid mass).[20]

The United States

Operational U.S. military satellites number about sixty. Most individual types of satellites are in LEO or GEO (see table 2-4 for unclassified satellite programs). However, in terms of total numbers, MEO is also heavily populated, due to the twenty-nine global positioning satellites now in that region. These provide navigation aid to military and civilian users; since 2000, they have provided both types of users with their positions to within about five meters.[21]

The U.S. military operates LEO satellites for ocean reconnaissance, weather forecasting, and ground imaging. The number of White Cloud ocean reconnaissance satellites that listen for emissions from ships probably has diminished with the effective end of the Soviet navy but may still total a dozen or more, deployed in groups of three at altitudes of roughly 1,000 kilometers.[22] The United States has two weather satellites, known as Defense Meteorological Satellite Program systems, in polar LEOs (they also carry gravity-measurement, or geodetic, sensors).

The United States also deploys probably half a dozen high-resolution imaging satellites in that LEO zone. They are of two principal types: radar imaging satellites, known as Lacrosse or Onyx systems, and optical imaging satellites, known as Keyhole systems, with the latest types designated KH-11 and KH-11 follow-on or advanced satellites. The Lacrosse radar satellites operate at roughly 600 to 700 kilometers above Earth, are capable of effective operations in all types of weather, and produce images with sufficient clarity to distinguish objects one to three meters apart. The KH satellites are capable of nighttime as well as daytime observations, by virtue of their abilities to monitor infrared as well as visual frequencies. They acquire information digitally and transmit it nearly instantaneously to ground stations. Their mirrors are nearly three meters in diameter, and they move in slightly elliptical orbits ranging from about 250 kilometers at perigee (point of closest approach to Earth) to 400 kilometers or more at apogee. Ground resolutions are as good as roughly 15 centimeters (6 inches) or even less under normal daylight conditions.[23] They can take images about 100 miles to either side of their orbital trajectories, allowing a fairly wide field of view.[24] They do not work well through clouds, however.

In GEO or near-GEOs, the United States deploys communications satellites, early-warning satellites for detecting ballistic

Table 2-4. Major U.S. Military Satellite Systems, as of 2003[a]

Satellite	Mission/capabilities	First launch	Number in con-stellation	Orbit altitude (miles)	Weight (lbs)
Advanced extremely high frequency satellite communication system (AEHF)	Successor to MILSTAR; provides world-wide command and control communications at five times the capacity, but in a smaller unit	2007, planned	4	22,300	13,000
Defense Meteorological Satellite Program (DMSP)	Collects air, land, sea, and space environmental data to support worldwide strategic and tactical military operations	1962	2	600	2,545
Defense satellite communications system (DSCS) III	Nuclear-hardened and jam-proof spacecraft used to transmit high-priority command and control messages	1982	5 (10 on orbit)	22,000+	2,580–2,716
Defense Support Program (DSP)	Provides early warning of missile launch by detecting booster plume	1970	Classified	22,000+	5,000
Global broadcast system (GBS)	Wideband communication system to provide digital multimedia data to warfighters	1998	3	23,230	Uses a variety of systems

System	Description	Year	Number	Altitude (miles)	Weight (pounds)
Global positioning system (GPS)	Precise location anywhere on Earth	1978	28	10,900	2,174; 2,370
MILSTAR Satellite Communications System (MILSTAR)	Joint communications satellite that provides secure jam-resistant communications for essential wartime needs	1994	5	22,300	10,000
Polar military satellite communications (Polar MILSAT-COM)	EHF payload on a host satellite to provide a cheaper alternative to MILSTAR for polar communications	1997	3 (1 on orbit)	25,300	470
Space-based infrared system (SBIRS)	Advanced surveillance system for missile defense, missile warning, battlespace characterization, and intelligence	FY 2007, planned	High (none on orbit)	GEO, elliptical, and low	Unknown
UHF follow-on satellite (UFO)	Secure anti-jam communications	1993	4 (9 on orbit)	22,300	2,600–3,400
Wideband gap-filler system (WGS)	High-data-rate satellite broadcast system designed to bridge the communications gap between current systems and an advanced wideband system	FY 2005, planned	3	GEO	13,000

Source: Mehuron, "2003 Space Almanac," pp. 41–42.

a. Classified systems are not shown.

missile launch, and signals-intelligence satellites for listening to other countries' communications or the emissions of their electronics systems, such as surface-to-air radars. Specifically, in the communications domain it has three global broadcast system (GBS) satellites for high-data-rate wideband communications; four Navy fleetsatcom (FLTSAT) satellites for communications primarily with ships; many Air Force satcom (AFSATCOM) packages on various hosts (including GPS satellites in MEO) for tactical communcations; roughly five functional defense satellite communications system (DSCS) satellites; and four MILSTAR (Military Strategic and Tactical Relay) satellites hardened against nuclear effects and jamming for critical communications. It also has about three defense support program (DSP) satellites for early warning of ballistic missile launch (as with most of its imaging satellites, exact numbers are classified).

Finally, the United States fields a handful of signals-intelligence satellites in GEO, though like the Lacrosse, Keyhole, White Cloud, and DSP systems, their exact number is classified.[25] The signals-intelligence satellites include the Magnum, with an antenna reportedly 200 meters wide for eavesdropping on communications.[26] Jumpseat satellites, flying elongated orbits, were developed to listen into communications from northern parts of the Soviet Union. That region could not be monitored by Magnum or its predecessor satellites in GEO because the curvature of Earth prevented the signals from reaching them.[27]

In a survey of American satellite capabilities, it would be a mistake to overlook the U.S. military's use of commercial satellites. The Pentagon used more commercial than military bandwidth during the 2003 war in Iraq. As noted, under an April 2003 policy directive, the Bush administration directed all government agencies, including the military, to look first to the commercial sector to meet their imaging needs.[28] The Pentagon has plans to spend about $100 million a year on commercial imagery with resolutions of one meter or better.[29]

The United States puts its military payloads into orbit from launch facilities at Cape Canaveral in Florida and Vandenberg Air Force Base in California. It also operates a half dozen smaller sites for some civilian payloads (see table 2-5). Its main rockets include the Atlas, Delta, and Titan families, as well as Pegasus and Athena rockets. The space shuttle has carried out some military functions as well, though it remains to be seen how much it will do into the future, in light of the 2003 *Columbia* tragedy and the planned retirement of the shuttle fleet by 2010. The U.S. military's future rocket requirements are to be met by the Atlas V and Delta IV, the two specific systems being developed under the so-called evolved expendable launch vehicle (EELV) program.

Satellites require ground stations to monitor and remotely maintain them, to adjust their trajectories, and to download data. Major ground control stations include Onizuka Air Force Base near San Francisco, Schriever Air Force Base near Colorado Springs, Vandenberg Air Force Base in California, and Pine Gap in Australia.[30] Other facilities important for one satellite constellation or another include Fort Belvoir, Virginia; Fort Meade, Maryland; Menwith Hill, United Kingdom; Guam; Diego Garcia in the Indian Ocean; Adak, Alaska; Winter Harbor, Maine; Kaena Point, Hawaii; Landstuhl, Germany; and Norfolk, Virginia.[31] Additional Space Command sites around the world include Thule, Greenland; Kwajalein Atoll in the Pacific; Ascension Island and Antigua in the Atlantic; Ramey in the Caribbean; Moron, Spain; and Misawa, Japan.[32] Up to half a dozen mobile terminals were developed during the cold war, primarily for receiving warning data about ballistic missile launch and communicating with strategic forces.[33]

Table 2-5. U.S. Space Launch Sites

Site	Civil or military	Location	Missions and operations	Number of launches, 1957–2002	Launch vehicles
Cape Canaveral AFS	Military	Florida, 28.5° N, 80° W: USAF East Coast launch site	Launches civil, military, and commercial satellites into GEO via ELVs and serves as ballistic missile defense (BMD) test facility	583	Athena I, II; Atlas II, III, V; Delta II, III, IV; Titan IV
Vandenberg AFB	Military	California, 35° N, 121° W: USAF West Coast launch site	Launches weather, remote sensing, navigation, communications, and reconnaissance satellites into polar orbits and serves as Defense ICBM and BMD test facility	620	Athena I; Atlas II, III, V; Delta II, III, IV; Pegasus; Taurus; Titan II, IV
Alaska Spaceport	Civilian	Alaska, 57.5° N, 153° W	Conducts polar and near-polar launches of communications, remote sensing, and scientific satellites up to 8,000 lbs.	6	Athena I, suborbital
Florida Space Authority	Civilian	Florida, 28.5° N, 80° W	Various launch complexes and support facilities for the USAF and Kennedy Space Center	6	Athena I, II; Minotaur; Minuteman III; Taurus; Terrier

John F. Kennedy Space Center	Civilian	Florida, 28° N, 80° W	NASA's primary launch base for the space shuttle	133	Pegasus, space shuttle, Taurus
Sea Launch	Civilian	Pacific Ocean, 0° N, 154° W	Heavy lift geosynchronous transfer orbit (GTO) launch services for commercial customers	8	Zenit-3SL
Spaceport Systems International, L.P.	Civilian	California 34.7° N, 120.46° W	Polar and near-polar launches from Vandenberg and payload processing facility	2	MM II—Delta III class
Virginia Space Flight Center	Civilian	Virginia, 38° N, 76° W	Launch facility for access to inclined and sun-synchronous orbits; recovery support, storage, and processing facilities	13 (since 1995)	Athena I, II; Black Brant; Lockheed Martin HYSR; Minotaur; Orion; Pegasus; Taurus; Terrier
Wallops Flight Facility	Civilian	Virginia 38° N, 76° W	East Coast launch site for Orbital Sciences' Pegasus and Defense missions	30	Pegasus

Source: Mehuron, "2003 Space Almanac," p. 34.

As for tracking objects in space, today most countries conduct space surveillance using telescopes and radar systems on the ground. Only the United States has a system providing some semblance of global coverage (though its southern hemisphere capabilities are quite limited). Its monitoring assets are located in Hawaii, Florida, Massachusetts, England, Diego Garcia, and Japan (see figure 2-3).[34]

A brief word on military organization is in order, though this complex and important subject is not a main focus of this book. The U.S. Air Force conducts and funds 80 to 90 percent of the overall American military effort on space and provides a comparable fraction of the uniformed personnel specializing in military space activities (about 35,000).[35] The Air Force's efforts are directed under the umbrella of the 14th Air Force at Vandenberg Air Force Base (AFB), California—which is one of the two constituent parts of Air Force Space Command in Colorado, the other part focusing on nuclear-tipped missiles. For satellite operations and ground control, the 14th Air Force includes the 21st Space Wing at Peterson AFB, Colorado, the 50th Space Wing based at Schriever AFB, Colorado, and the 460th Air Base Wing at Buckley AFB, Colorado. For space launch, the 14th Air Force employs the 30th Space Wing at Vandenberg, largely for putting satellites into polar orbit and for long-range missile tests, as well as the 45th Space Wing at Patrick AFB, Florida, largely for launching GEO satellites from Cape Canaveral. The other military services also have space-related commands, notably the Army's Space and Missile Defense Command, based in Virginia with an operations headquarters at Peterson AFB in Colorado and a research headquarters in Alabama. But these are smaller operations, making use of fewer resources.

In the future, the United States will naturally improve a number of its satellite capabilities. Specific programs involve additional MILSTAR satellites, a UHF-band follow-on communications system, three advanced EHF satellites, at least two more Defense Satellite Communications System (DSCS) spacecraft, three Wideband Gapfiller satellites, and four Mobile User Objective System (MUOS) spacecraft.[36] (EHF signals include frequencies from 30,000 to 3,000,000 megahertz; SHF covers the range from 3,000 to 30,000; UHF from 300 to 3,000; VHF from 30 to 300; and HF from 3 to 30. In each case, the letters HF stand for high frequency, with the other letters representing the words extremely, super, ultra, and very.)[37] In addition to this heavy focus on communications systems, starting around 2004, GPS satellites will be gradually upgraded to a more robust, jam-resistant variety with a dedicated higher-power military frequency. The United States will also modernize its early warning capabilities in GEO by deploying the so-called SBIRS-High (space-based infrared satellites at higher orbits) system. And it will purchase low-altitude satellites in the Space Tracking and Surveillance System (STSS) to track missile warheads.[38] The total price tag for all of these systems is estimated at roughly $60 billion.[39]

The United States also will surely improve its imaging and signals-intelligence systems, currently in LEO and GEO, respectively, though data on these classified activities is difficult to come by. The broad name for the project under which imaging capabilities will be developed and deployed is the Future Imagery Architecture. The FIA research and development program was purportedly to cost $4.5 billion in all, but it has faced major challenges and setbacks. Little more is known about the project to date. The United States may try to develop a constellation of radar satellites in space for purposes of continuously

Figure 2-3. Locations of U.S. Ground-Based Space Surveillance Sensors, Past and Present

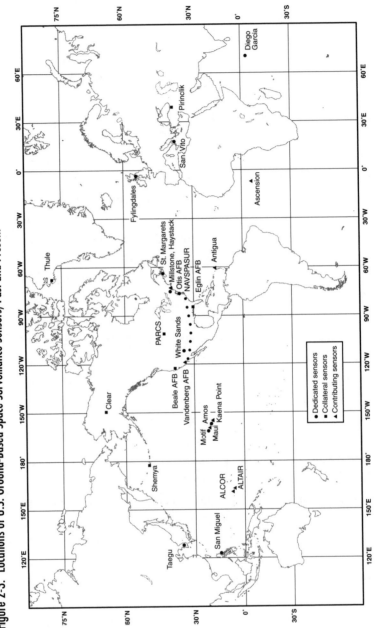

Source: Paul B. Stares, *Space and National Security* (Brookings, 1987), p. 206.

tracking and targeting objects on Earth. Toward this goal, a program known as Discoverer II was initiated in the 1990s; the idea was to put twenty-four to forty-eight satellites into LEO to provide nearly continuous coverage.[40] But difficulty in making any progress toward cheaper satellites and other technical challenges recently led to the program's cancellation. (Such assets might cost $1 billion apiece today, and the constellation would appear unaffordable unless the unit price could be brought down to about $100 million.) The Pentagon's 2005 budget proposed $327 million in funding for a new space-based radar program.

Over the longer term, the United States is clearly interested in considering development and deployment of other possible systems, such as space-based lasers (SBLs) or kinetic-kill vehicles, in LEO for purposes of ballistic missile defense. And any ground attack systems for attacking objects on Earth from space, as distant a prospect as such capabilities may now be, would presumably be in the LEO zone as well.[41]

Russia

Although it has clearly fallen from its superpower status, Russia remains the world's second space power by most meaningful measures. Its total space budget is roughly that of India, and as a result of the dissolution of the former Soviet Union it has lost direct control over a number of facilities (such as directed-energy test beds at Dushanbe in Tajikistan and Sary Shagan in Kazakhstan).[42] But it continues to put satellites into space at an impressive pace, averaging more than twenty-five launches a year in recent times, in contrast to a U.S. level of around thirty. It does so using at least eight different families of launch vehicles of many sizes and payloads, including Molniya,

Soyuz, Cosmos, Shtil, and Start variants. It operates five of the world's twenty-seven major launch sites (see table 2-6).[43] Russia's manned space program also continues. In recent years, it has maintained a typical flight schedule of two launches with three to six cosmonauts per year.

Russia has more than forty working military satellites by recent estimates, close in quantity to the United States. They run the gamut from communications and navigation assets to early-warning satellites to electronic intelligence devices.[44] But while Russia rivals the United States in numbers of satellites, the equality ends there. For example, Russia's GLONASS navigation system, designed to have nine to twelve satellites and provide just moderately accurate locational information in a best case, had eight working satellites at last count. Only half of Russia's early-warning satellites were still operational in 2000.[45] It reportedly had only one imaging satellite in service in the late 1990s and may have experienced a period when none was functional. Overall, perhaps two-thirds of its satellites were beyond their planned service lives by the end of the 1990s, and little has improved since.[46]

China

China has more than thirty satellites in orbit. It has been increasingly active, with up to half a dozen launches a year in recent times. It operates three launch sites and is an increasingly popular low-cost provider of orbiting services. It also is working on a manned space program, run by the People's Liberation Army (PLA), and put its first astronaut into space in 2003. It also hopes to put an unmanned vehicle on the moon by 2010.

Most of China's satellites are at least nominally civilian, as opposed to military, assets. For example, it fields five civilian Earth observation satellites in LEO (one in conjunction with

Table 2-6. Orbital Launch Sites Worldwide

Site	Owner	Number of launches, 1957–2002
Plesetsk	Russia	1,535
Tyuratam/Baikonur, Kazahkstan	Russia	1,190
Vandenberg AFB	United States	620
Cape Canaveral AFS	United States	583
Kourou, French Guiana	European Space Agency	166
John F. Kennedy Space Center	United States	133
Kapustin Yar	Russia	101
Tanegashima	Japan	35
Xichang	China	33
Kagoshima	Japan	30
Shuang Cheng-tsu/ Jiuquan	China	30
Wallops Flight Facility	United States	30
Edwards AFB	United States	20
Sriharikota	India	16
Taiyuan	China	16
Indian Ocean Platform	United States	9
Pacific Ocean Platform	Sea Launch	8
Palmachim	Israel	5
Hammaguir, Algeria	France	4
Svobodny	Russia	4
Woomera	Australia	4
Alcantara	Brazil	2
Barents Sea	Russia	1
Gando AB, Canary Islands	Spain	1
Kodiak	United States	1
Kwejalein, Marshall Islands	United States	1
Musudan ri	North Korea	1
Total		4,579

Source: Mehuron, "2003 Space Almanac," p. 26.

Brazil) and some thirteen civilian communications satellites in GEO (some of which may no longer be functional), as well as two scientific devices in LEO, one weather satellite in LEO, and three navigation devices.[47]

China uses a half dozen space launch vehicles in the Long March series. Most are three-stage rockets. Their payloads range from 2,000 to 10,000 pounds per launch.[48] An improved family of liquid-fueled rockets is being developed. One variant is expected to have, among other features, the capacity to lift 24,000 pounds to LEO.[49]

China is improving its satellite and space capabilities with vigor. It appears to be interested in developing imaging satellites based on electro-optical capabilities, synthetic aperture radar, and other technologies. Its Ziyuan imaging satellites, planned in conjunction with Brazil, would have real-time communications systems to get data to the ground quickly, as would be needed for tracking mobile military targets. It is also cooperating with Russia on a number of space programs, possibly including satellite reconnaissance technology.[50] And according to Desmond Ball, it is quite possibly making progress on electronic intelligence satellites.

Europe and Japan

A number of other major industrial countries field substantial numbers of satellites; in addition, the European Space Agency deploys thirty-two. Dozens are for communications (though only four were dedicated to military communications purposes as of this writing). Half a dozen are for Earth observation, including France's two SPOT satellites and its two Helios military satellites; radar imaging may follow.[51] About a dozen are for scientific purposes; and two are for weather forecasting.[52] The European Space Agency and the European Com-

mission are now completing contracting for the Galileo navigation system, which is intended as a GPS-like constellation of
thirty navigation satellites, costing about 3.5 billion euros
(including ground stations). The ambitious schedule would
have initial service begin in 2008.[53] It may be somewhat redundant but promises also to be more robust than GPS and provide
broader and more continuous coverage—meaning that there is
little point in the United States trying to block it. Rather, the
United States needs to stay on the ball in its own plans for GPS
modernization if it is to compete successfully.[54]

Four of the world's major launch sites are in Europe or are run
by Europeans. And the European Space Agency has been launching about ten rockets a year into space in recent times. Europe's
rockets include France's Ariane family, Italy's Vega rocket, and
the Sea Launch Zenit (Sea Launch is a company made up of
American, Russian, Ukrainian, and Norwegian partners).[55]

Japan tends to average one or two satellite launches a year. It
operates two launch sites. Its current launch vehicle is the H-2A,
a smaller variant of the ill-fated H-2, which suffered two launch
failures and never became a dependable system despite years of
investment and effort. Japan's orbiting assets include up to sixteen GEO communications satellites, three scientific satellites,
and a weather satellite. In March 2003, it orbited a pair of imaging satellites, one optical and one radar, with resolutions of
roughly one meter and one to three meters, respectively.[56]

Other Countries and Companies

Beside the countries discussed above, about twenty-five other
states have at least one or two satellites in space. Several countries, including Canada, India, Israel, Thailand, South Korea,
Brazil, and Argentina, own reconnaissance satellites. They operate them largely for environmental and economic and planning

purposes, but sometimes for military missions as well. Large countries with numerous remote regions, such as India, Canada, and Indonesia, as well as Russia and China, tend to see satellite communications as important for civilian purposes. Such capabilities clearly could have military applicability as well. Israel does not need a satellite to connect geographically separated regions but is planning to build a dedicated military communications satellite.[57] Israel is also capable of producing images with half-meter resolution and is considering selling some of them to India.[58]

India operates its own launch site, as do Australia, Israel, Brazil, Spain, Ukraine, and North Korea. Major families of boosters include India's GSLV, Brazil's VLS-1, and Israel's LK-1 and Shavit rocket systems.[59]

Commercial firms such as the French-British SPOT and the American LANDSAT have provided Earth images for many years. LANDSAT began with a satellite providing eighty-meter resolution in 1972; by 1982, its latest model had improved that figure to thirty meters. SPOT images have resolutions of about ten meters. The capacity to produce images with even these levels of resolution was enough to cause concern during Operation Desert Storm in 1991 (they could have revealed preparations for the famous "left hook" around Iraqi forces, among other things). No images were provided to the Iraqis, but the concern clearly remains for possible future wars.

Technology and commercial trends are clearly giving a growing number of companies, many not American, the capability of producing high-resolution images for sale. Eleven such companies are expected to be in the market soon, including firms from Canada, India, Israel, Russia, and possibly China. Recent U.S. commercial efforts include Space Imaging, with its Ikonos satellite; Earth Watch, with its Quickbird; and Orbital Sciences, with its OrbView-3 and OrbView-4. The Ikonos satellite has a resolution slightly better than one meter with black-and-white images; a satellite company known as Orbimage anticipates one-meter black-and-white resolution soon, as well.[60] Space Imaging is considering orbiting a satellite with twice as fine resolution within a couple of years.[61]

Other countries and companies are not too far behind. Relevant private ventures include the International Telecom Satellite Organization, Globalstar, Orbcomm, the European Telecom Satellite Organization, the International Maritime Organization, the Arab Satellite Communications Organization, and the Asia Satellite Telecom Company. As one example, a constellation of eight Eros imaging satellites, owned in part by the Israeli government, with one- to two-meter resolution, is expected to be in orbit soon.[62]

Conclusion

Satellite technology is complex and challenging, as are rockets for putting satellites into space. But capabilities are becoming widespread nonetheless. The rapid growth in space-related business seen during much of the 1990s has slowed, but trends are still upward and can be expected to remain that way. Space is becoming somewhat crowded, not only with satellites but also with debris from rocket parts and old satellites. Regardless of how one views it, the recent period during which the United States has dominated, and indeed almost monopolized, the military use of space seems unlikely to endure—though no country is anywhere close to approaching American capabilities in either commercial or military terms.

CHAPTER THREE

CURRENT THREATS AND TECHNOLOGY TRENDS

If space-related technologies could be frozen in place in their current state, the United States would be in a fortunate position. It dominates the use of outer space for military purposes today. Russia's capabilities have declined, to the point where its weakness may be of greater concern than its strength, given the lack of dependable early-warning satellites for ballistic missile launch. China's assets remain rudimentary, as do those of America's other potential rivals (or current enemies). The United States is able to use satellites for a wide range of missions, including not only traditional reconnaissance and early-warning purposes but also real-time targeting and data distribution in warfare. Although it hopes to develop space-based missile defense assets someday, the present need for such capabilities is generally rather limited, and ground-based systems increasingly provide some protection, in any event. More exotic capabilities, such as space-to-Earth kinetic rods, airplanes that would bounce along the top of the atmosphere, or intercontinental artillery, are not of pressing need given existing U.S. capabilities for projecting power anywhere in the world. On the whole, the current configuration of global space technologies and assets is highly desirable from an American perspective and unlikely to improve much further, if at all. Deterioration in the U.S. position seems more likely.

Of course, it is not possible to freeze progress in technology, nor to stop the continued dissemination of technologies already available. So American policymakers will have to adapt. This chapter provides a prognosis on some key areas of space-related technology over the coming ten to fifteen years to provide a foundation for consideration of various policy options in chapter 6. It also underscores current vulnerabilities of American military and commercial satellites, including some that are probably underappreciated by policymakers (see table 3-1).

The chapter focuses on a few key technologies, including high-powered lasers, launch vehicles, and microsatellites. The emphasis is on fundamental trends in physics and engineering, rather than on possible adaptations of existing technologies (such as solar-powered stations or nuclear reactors in space).[1] In general, the emphasis is on the basic question of what will become possible in space, rather than what might save money or offer new conveniences. In addition, one old but sometimes forgotten threat, the possibility of nuclear explosions in space, is also considered.

One general theme about future technology is that, despite the tendency of military strategists to rave about defense transformation and a coming revolution in military affairs, many satellite development programs are currently advancing more slowly than once hoped. Leaving aside fundamental constraints of the laws of physics, immediate engineering challenges are making it harder than expected to develop systems that are generally believed to be within reach. For example, the nation's next generation of imagery satellites, known as the Future Imagery Architecture, has recently been delayed by more than a year and grown by more than $3 billion in cost. Problems also afflict next-generation global positioning system satellites, space-based infrared satellites at higher and lower orbits (SBIRS-high and SBIRS-low), and communications systems (such as the advanced extremely high-frequency satellite system).[2] Cheaper and/or reusable launchers are proving hard to develop as well, as discussed below. Most futuristic technologies remain just that.

But another competing theme is that some areas of technology are indeed changing in important ways. Chapter 1 has documented the increasing spread of satellite capabilities. As time advances, these can be expected to provide more and more countries with the ability to mimic the United States in real-time targeting of military assets for tactical warfighting purposes. Other countries will surely remain behind the American military in their absolute capabilities—data rates per second, raw processing power, wherewithal for global operations, ability to find and destroy a wide range of military targets with high confidence. But they may not need to compete on equal footing with the United States to cause it great concern. For example, for China, Iran, or another regional power to threaten shipping near its shores, including possibly U.S. Navy assets, may only require an intermittent ability to strike at a small number of large assets. Such developments are not imminent, and are not reason enough for a U.S. decision to develop antisatellite weapons in

Table 3-1. U.S. Space System Survivability Measures, up to 1987[a]

Mission or program	Nuclear hardening	Maneuver-ability	Laser hardening	Satellite redundancy	Anti-jamming capability	Link resistance to nuclear effects	Satellite autonomy	Ground station redundancy	Overall survivability rating
Reconnaissance	L(P)	n.a.	n.a.	L	n.a.	L	n.a.	L	Low
Early warning	L	M(P)	L(P)	M	L(P)	L(P)	L(P)	L(P)	Low to medium (P)
Communication									
DSCS III	L	H	n.a.	L	H	M	M	M	Medium
FLTSATCOM	L	M	N	L	L	L	L	L	Low
Leasat	L	n.a.	N	L	L	L	L	M	Low
AFSATCOM	—	M	—	—	L(P)	L	—	—	Low
SDS	L	n.a.	N	L	L	L	L	L	Low
NATO III	L	L	N	L	L	L	L	L	Low
MILSTAR	H	H	H	M	H	H	H	H	High
Navigation									
Transit-Nova	L	M	N	M	n.a.	L	L	L	Low
Navstar GPS	n.a.	H	H	H	H	H	H	H	High
Nuclear explosion detection (NDS)	—	H	n.a.	H	n.a.	L	H	H	Medium to high
Meteorology (DMSP)	L	H	L(P)	L	L(P)	L(P)	L(P)	L(P)	Low to medium

Source: Paul B. Stares, *Space and National Security* (Brookings, 1987), p. 198.

L = limited protection; M = medium protection; H = high protection; N = no protection; (P) = planned improvement; n.a. = not available.

a. After the cold war ended, less emphasis was placed on survivability, so many of these estimates are believed to remain correct today.

the foreseeable future, as is argued in later chapters. But these trends bear watching, nonetheless.

New technologies will also be developed, and the United States may not always lead the way. Three major possibilities stand out. One is in the domain of high-energy lasers, capable of damaging many unhardened and unprotected objects in low-Earth orbit even if situated on the ground and even if lacking adaptive optics to compensate for atmospheric distortion. A second, perhaps even more important, development pertains to microsatellites. Modern computing and other technological improvements are making it possible for small satellites weighing just dozens of kilograms to maneuver and operate autonomously in space. They cannot necessarily maneuver extensively on their own or provide multifaceted capabilities. But they may soon be capable of harassing or destroying larger satellites at virtually any altitude—and of being positioned to do so clandestinely. U.S. space tracking assets today could watch any rocket carrying such microsatellites into space, but they might not detect small objects making up part of the payload, and might also lose track of a small maneuvering object over time. Finally, the improvement of "hit-to-kill" ballistic missile defense technologies against long-range missiles will provide latent capabilities against low-altitude satellites as well, since any weapon capable of intercepting a ballistic missile warhead in space should be capable of intercepting a satellite at comparable altitudes. Such satellites would be moving at only slightly greater speeds than missile warheads, and along relatively similar and even more predictable trajectories. In addition, the targeting and communications system used to guide a ballistic missile interceptor to its target could be easily modified to accept data from satellite surveillance and tracking stations (see figure 3-1).

Figure 3-1. ASAT Intercept Schemes[a]

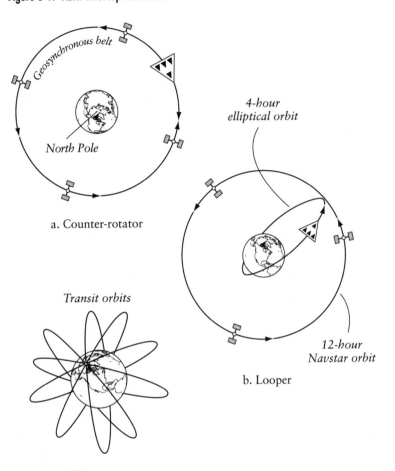

a. Counter-rotator

b. Looper

c. Direct ascent

Source: Ashton Carter, "Satellites and Anti-Satellites: The Limits of the Possible," *International Security*, vol. 10, no. 4 (Spring 1986), p. 83.

a. ASAT intercept schemes aim to be able to attack all the satellites of a constellation with a specific set of orbital maneuvers. The counter-rotator (panel a) travels the GEO belt in the "wrong" direction, attacking all satellites within twelve hours. The looper (panel b) climbs to semisynchronous orbit every four hours to pick off one of the Navstar GPS satellites, which are phased every four hours in a twelve-hour orbit. A battery of direct ascent ASAT interceptors based at the North Pole (panel c) could attack all satellites in polar LEO in less than two hours.

Underappreciated Threats:
Nuclear Detonations and Microwave Weapons

Nuclear weapons are an effective means of targeting satellites. They are often carried by ballistic missiles with guidance systems that could easily be reprogrammed to detonate at a point in space; if it was known when a given satellite would pass near that point, close-proximity intercept would not be difficult to achieve (even without testing for that purpose). Nuclear weapons are, of course, lethal from close range, even against hardened satellites. Any country with nuclear weapons and even relatively short-range ballistic missiles might be able to generate this type of threat, since low-Earth orbit is so near and this type of attack does not require great accuracy or finesse.

Some argue that adversaries would desist from using nuclear weapons in space out of fear of retaliation. It is true, certainly, that this would be a provocative action with considerable potential for inciting some type of escalation from the United States. But the assumption that an enemy would be deterred for that reason is unconvincing and too optimistic. What better way to use nuclear weapons than to destroy a key military capability of an enemy country without killing any of its population? The United States could threaten nuclear retaliation after such an attack, but it is far from clear that such a threat would be credible—or even appropriate. And an enemy might feel it had little to lose anyway, if the United States was already bent on regime overthrow as its ultimate objective in the war in question. On balance, this concern is considerably more serious than many appreciate.

LEO satellites are clearly the most vulnerable to nuclear attack, though unshielded commercial satellites as far up as GEO could be vulnerable to disruption from nuclear explosions at lower altitudes.[3] But in LEO, even hardened military satellites are vulnerable. LEO satellites often are few in number in any given constellation. More important, they can be roughly 100 times closer to Earth than a GEO asset, at least when passing directly overhead. Even SCUD-class missiles are capable of reaching them, if a nuclear warhead is available small enough to fit atop the SCUD. (By contrast, an ICBM could only reach GEO if its normal payload was reduced by roughly a factor of five, given the physics of rocket flight.)

Nuclear bursts can harm satellites in several ways. One is direct destruction through proximity to the blast (especially due to x-rays). They can destroy satellites at several hundreds of kilometers from a detonation point, depending most importantly on the hardening of a satellite and to a lesser extent the size of the explosion. A second possible means of damage is through x-rays created by system-generated electromagnetic pulse, for example from a burst at an altitude of 100 to 150 kilometers (which could damage or destroy 5 to 10 percent of a constellation of LEO satellites, specifically those within line of sight of the detonation).[4] Either one of these mechanisms can damage satellites thousands of kilometers away; some satellites could be affected at distances of 20,000 to 30,000 kilometers (by a one-megaton blast, assuming limited shielding of the satellite).[5]

A third mechanism by which nuclear weapons could damage or destroy satellites is a more gradual effect. It results from the so-called pumping of the Van Allen radiation belts by any explosion, particularly one crossing the 50 kiloton threshold, anywhere from roughly one hundred to several hundred kilometers in altitude.[6] Pumping of the Van Allen belts results when fission products emit protons and electrons that are trapped by Earth's magnetic fields, striking satellites as they repeatedly orbit through these regions in the ensuing days and months. For example, a 1962 U.S. megaton-class test at 400 kilometers over Johnson Island in the Pacific, known as Starfish, destroyed some seven satellites in seven months and continued to affect the Van Allen belts until the early 1970s.[7] A fledgling nuclear power might not have an explosive device approaching this yield, but even smaller explosions would have deleterious effects of this type. This mechanism is of greatest concern for unshielded satellites in LEO; some military satellites in LEO, and at higher altitudes within Van Allen belts (such as GPS satellites), are generally hardened and more resilient.[8] Typically, unhardened LEO satellites with expected lifetimes of five to fifteen years might last only a few months or less under such conditions (see figure 3-2).[9]

These vulnerabilities are worrisome. They raise the question of whether hardening requirements might be placed on U.S. commercial satellites, with the government possibly subsidizing the added costs, which often total about 2 to 3 percent of the satellite's value.[10] But even more to the point, they raise the question of whether current U.S. military satellites, many of which are probably not hardened very well against nuclear effects, need better protection. (They were not even hardened that well during the cold war, and the degree of hardness on subsequent satellites has probably declined.)

Those not convinced that a potential adversary would use nuclear weapons against U.S. satellites should still worry about high-powered microwave weapons. These explosive-driven devices have the ability to generate very high power levels in an intense burst. Large devices could be lethal to electronics out to a range of many kilometers, perhaps dozens or even more; smaller devices, such as those that could be placed on microsats, could be effective at distances of perhaps tens to hundreds of meters. The physics behind these weapons is not complicated,

Figure 3-2. Estimated Effects of Low-Yield, High-Altitude Nuclear Detonations on the Service Lives of Selected LEO Satellite Constellations[a]

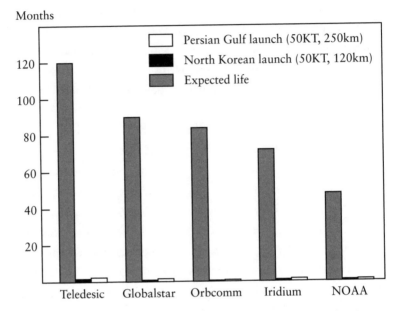

Months

Legend:
- Persian Gulf launch (50KT, 250km)
- North Korean launch (50KT, 120km)
- Expected life

Categories: Teledesic, Globalstar, Orbcomm, Iridium, NOAA

Source: Barry D. Watts, *The Military Uses of Space: A Diagnostic Assessment* (Washington: Center for Strategic and Budgetary Assessments, 2001), p. 99.

a. System hardness is assumed to be twice that needed for the natural environment. Satellite constellations are as follows: Teledesic, 1,350 km, 98° inclination; Globalstar, 1,414 km, 52° inclination; Orbcomm, 775 km, 45° inclination; Iridium, 780 km, 84.6° inclination; NOAA, 850 km, 99° inclination.

and the engineering requirements are potentially within the reach of many countries. Hardening against nuclear weapons is not the same as hardening against these types of microwave weapons, due to differences in wavelengths and intensities of radiation. But there could be economies of scale and effort in carrying out both types of hardening as part of the overall design or redesign of a given satellite.[11]

High-Energy Lasers on the Ground and in Airplanes

What about technology areas that are changing significantly? High-energy lasers are a good place to start. Those with potential as long-range weapons would probably be chemical lasers that would destroy their targets by heating them with continuous waves of infrared radiation. Unlike nuclear weapons, which generate intense x-rays, or pulsed lasers, they would not produce sufficiently high-frequency rays to destroy through the sheer penetrative and disruptive power of the photons or through the physical jolt delivered to the system as a short impulse. But many chemical lasers could have some effectiveness against unprotected objects in LEO when they flew overhead; more powerful devices could damage satellites further away. Such lasers are usually able to convert about 20 to 30 percent of the energy released by chemical reactions into laser power.[12]

To damage a soft target like paper or human skin, a total dose of about one Joule per square centimeter is required (a Joule is a watt of power applied for a second). Wood is generally damaged after receiving about ten Joules per square centimeter; metal, after 100 Joules per square centimeter. The type of target usually envisioned for high-energy laser weapons today, for example, the metal making up the skin of a SCUD missile, might be damaged after receiving 1,000 Joules per square centimeter.[13] By contrast, many satellites could apparently be damaged after receiving as few as ten Joules per square centimeter, assuming a pulse lasting several seconds, according to a 1995 Air Force scientific advisory study.[14] Other public estimates are in the range of fifty Joules per square centimeter, through overheating of a satellite body or solar panels—though 1,000 Joules per square centimeter might be needed to effect a quick, "brute force" kill of individual components.[15] Much depends on the specific characteristics of the satellites.[16] The main point is that satellites can be much easier to damage or destroy than SCUDs, meaning that they could be targeted from much longer distances.

Apart from chemical lasers, the two other types of laser technology that could most plausibly be used as weapons in the future are the free-electron laser and solid-state laser. But both are far removed from generating the types of power outputs required for the missile defense or antisatellite job. The first, which produces radiation by sending very fast electrons through a magnetic field, is about a factor of one thousand too weak at present. The second, which excites molecules inside a crystal solid until they can be induced to return to a lower energy state in unison, releasing radiation in the process, is now about a factor of one hundred too weak (at roughly 10 kilowatts).[17] Prospects for greatly increasing the intensity of solid-state lasers, in particular, are mediocre, given the difficulty of venting the heat created by the lasing process.[18] Moreover, since some 70 percent of total high-energy laser funding has been directed to large demonstration projects in recent years, breakthroughs in these types of lasers seem relatively unlikely anytime soon; only a few tens of millions of dollars a year have been devoted to more basic research.[19] The laser defense industrial base has also lost much of its strength in recent years, and most of today's demonstration efforts involve projects begun in the 1980s, based on technologies conceived in the 1970s.[20]

But what may be a curse for laser science in general is a boon to one U.S. program in particular: the airborne laser, currently in a fairly advanced stage of development as a missile defense system. This system may have rudimentary capabilities against ballistic missiles fairly soon, even though it will probably not be fully deployable until late in the decade (if then). It could, with relatively modest modifications, have ASAT capabilities anytime thereafter, should the United States elect to proceed down that path.

The idea of high-energy chemical lasers dates back to the 1970s. Airborne laser experiments go back two decades, to the early 1980s, when an aircraft outfitted as an "airborne laboratory" used a laser to defeat several Sidewinder missiles launched in its vicinity. In this period the MIRACL was also built at White Sands, New Mexico. That laser uses deuterium fluoride fuel and operates at 3.8 microns wavelength at the megawatt level.

The ABL enjoys two major advantages over MIRACL. First, it is airborne, meaning it can fly and operate above the atmosphere's most dense region and above almost all clouds. Since Earth's atmosphere interferes with most kinds of visible and near-visible light, scattering or absorbing much of it, this is a great benefit. In addition, the infrared wavelength used by the airborne laser is less affected by whatever atmosphere it does encounter (a wavelength range of 0.5 to 1.5 microns is considered ideal; the ABL operates at 1.315 microns).[21]

Each ABL is actually designed to be a system of lasers. The main beam is a high-power system for destroying an enemy missile. Other lasers of lesser power on the aircraft are designed for targeting and tracking and to measure atmospheric conditions. The ABL is designed first and foremost to work against liquid-fueled short-range missiles, such as SCUDs. It was defined as a theater missile defense capability in the Clinton administration, though it could certainly be used against long-range liquid-fueled rockets as well. So the ABL is really a boost-phase intercept concept for use against liquid-fueled ballistic missiles, regardless of their range.

Whether the ABL would work against solid-fuel ICBMs or not is unclear. Its range might be diminished, because the basic concept of weakening the booster body enough to produce a catastrophic leak and explosion might not work as quickly against solid-fuel rockets. On the other hand, range might also be increased, because the laser might be able to dwell on the target longer than with a shorter range missile.[22] The Missile Defense Agency is contemplating tests of the ABL against ICBMs in mid-decade that could shed more light on this matter.[23]

The ABL uses hydrogen peroxide, potassium hydroxide, chlorine gas, and water as raw ingredients. A number of modules (six on the first test aircraft, fourteen eventually) will together produce a beam with a strength of about 1 million to 2 million watts (1–2 megawatts) and a beam roughly the size of a basketball at hundreds of kilometers' range. It is to operate on a modified 747 aircraft. At the 40,000 foot altitudes where it is designed to function in wartime, it can be assured a relatively clear shot at targets at comparable or higher altitudes, since clouds rarely rise to that height. Its maximum range against a short-range ballistic missile is estimated at up to several hundred kilometers; with a single payload of chemical fuel, it could fire about twenty shots, each lasting several seconds.[24]

The ABL almost surely will have latent antisatellite capabilities. It can be directed against targets above it, perhaps even directly above it. Its beam can be swiveled in the horizontal plane almost back to the aircraft's wings. The method for pointing the beam appears to allow just as wide a sweep in the vertical direction—though exact figures are classified, perhaps because of political sensitivity that others will see it as a possible ASAT. It makes sense that the ABL could damage certain classes of satellites in low-Earth orbit, given that they often fly over Earth at altitudes comparable to the expected lethal radius of the ABL. (Of course, satellites' orbits would only occasionally take them over a spot where an ABL was flying.) But the main issue with converting the ABL into an ASAT probably concerns target acquisition and tracking. At present, the ABL relies on hot rocket plumes for acquisition of the target; overhead satellites would not provide such a signature. Thus the ABL could not track and destroy a satellite unless its tracking sensors were first cued to the satellite's location by the space surveillance system. The United States has such a system already. Providing the necessary data links would require software changes and perhaps even more, but it would not require changes to the basic laser system of the ABL. That means it could probably be accomplished relatively quickly.

The airborne laser does seem likely to work, even if schedule delays may continue (plans for the first true flight test of the integrated system, originally planned for 2002 and then 2003, have continued to slip). The benefits of having a laser above most cloud cover and much of the atmosphere are considerable, so the United States can be expected to continue providing strong support for the ABL. (The Navy gave up on similar technology for ship defense because of propagation losses for a high-powered laser of 3.8 microns' wavelength at sea level; flying at 40,000 feet and using a shorter wavelength beam is a much different matter, however.)

The engineering challenges still remaining for the ABL are considerable—modifying a chemical laser concept developed in 1977 to create a powerful laser small and light enough to fit on an airplane, dealing with atmospheric distortions, keeping the laser on the target while flying, and so on. Yet most of these are on the path toward solution in one way or another. Solving them may take time, but is likely to prove feasible. And for this reason, a system with an inherent ASAT capability against low-flying satellites seems likely to be in the U.S. inventory by the second decade of the century.[25]

What could other countries do to exploit high-energy laser technology for space weapons applications? Russia has much of the necessary expertise but may lack sufficient financial resources (and motivation). India may try to develop a laser that can be used against targets in space over the next decade or so, but its prospects for success are not yet clear.[26]

China is the more interesting concern for the future, especially given its ongoing disagreements with the United States over Taiwan, and thus the potential for war. China is making some progress with various laser technologies. It is now believed to have low-energy laser countermeasures for deflecting antitank missiles, showing that it has solved some of the challenges in pointing lasers and keeping them fixed on their targets. The Pentagon also believes that it may have acquired (perhaps from Russia) high-energy laser technology that could be used in antisatellite operations. There are some reports that it has thought about atmospheric "thermal blooming," an effect caused by the passage of high-powered laser light through the atmosphere that leads to the distortion and weakening of a high-powered laser beam if not properly addressed. But that concern is not fleshed out in the Pentagon's latest report on China's military capabilities.

Some U.S. analysis on China's capabilities has gotten a bit carried away. For example, the 1999 Cox report suggested that Russia might help China develop nuclear-pumped lasers in space, an extraordinarily challenging technology that probably remains twenty-five years in the future even for the United States, should it ever choose to pursue such a capability.[27]

But the matter does not end there. It is doubtful that China, or for that matter any other country, could develop an airborne laser capability in the next ten to fifteen years. The juxtaposition of various technologies and the resources required for such a program are probably beyond its means. Yet China may soon have the inherent ability to produce a ground-based high-energy laser like the MIRACL, should it devote the very substantial resources and time needed to make such a program work—even if such a weapon might lack adaptive optics and other sophisticated features that would help concentrate its power.[28] The technology is now a couple of decades old. Placing such a system on the ground may not make for an ideal ballistic missile defense, given atmospheric effects and the fact that Earth's curvature would prevent the laser from striking most missiles during most of their trajectory. But ASAT operations are easier to contemplate, since in many scenarios one can wait for a clear day and for the target to fly overhead.

Space-Based Laser Concepts

Although they, too, have been discussed and investigated for decades, space-based laser concepts are much further from fruition than ground-based systems or the ABL. The Pentagon acknowledges that they are probably ideas for 2020 and beyond.

The space-based laser program is currently run by the Missile Defense Agency and the U.S. Air Force. It would employ a different type of chemical laser that makes use of hydrogen and fluorine to create hydrogen fluoride, resulting in infrared radiation at a wavelength of 2.7 microns. That is about twice the wavelength of the airborne laser and is less suitable for use within the atmosphere; given how strongly radiation at that wavelength is absorbed by water vapor, it would probably only penetrate down to 30,000 to 40,000 feet if directed into the atmosphere from space.[29] But against targets in space that disadvantage clearly would not matter. The fuels are light and relatively stable, so good for long-term storage in space.[30]

In the SBL, a large mirror with a diameter of at least four meters and perhaps as much as eight meters would be used to create a fine beam. The mirror would have to be extremely light. It would probably have to be furled up while being deployed and unfolded once in space. The laser would have a length of about twenty meters and weigh nearly twenty tons, according to current plans. The program's goal is to move toward a lethal demonstration of the system in orbit by 2012, but a constellation of a dozen or more satellites providing global coverage is probably at least two decades away, according to a 2001 Defense Science Board assessment and other sources.[31]

By contrast with the ABL, the space-based laser concept remains little more than a basic research and development concept in the Department of Defense. Current funding levels are in the range of tens of millions of dollars a year. If a demonstrator is built over the next decade or so, it is currently estimated that costs would total about $3 billion. To get to that point would require realistic ground testing, which would necessitate construction not only of the laser but of a test facility mimicking space—which means a large vacuum chamber.[32] Even if the basic laser package could be built and orbited, major improvements would be needed prior to building actual weapons, including development of deployable larger-scale optical systems, at least a fivefold increase in power relative to today's levels, and progress in jitter control that would stabilize the laser and keep the beam fixed on its target.[33]

Each SBL would essentially be a combination of three extremely complex technologies: the laser itself, the power source for the laser, and the equivalent of a space telescope to direct the beam. Integrating these elements may be no harder than in the airborne laser. Indeed, a space-based laser would not have to deal with any atmospheric distortion of its beam, as does the ABL. But in other ways, the challenge associated with the SBL is much greater. It is already proving difficult to put lasers on aircraft with payloads of 100,000 pounds or more; it is far harder to put them into space with rockets each capable of lifting payloads less than half that weight. Even if high-powered lasers, space telescopes, and large fuel payloads could be individually orbited, assembling them in space and making them work in that environment for the purposes of missile defense or antisatellite operations is a far more challenging proposition. (No launcher system is presently being developed with the requisite capacity to launch a space-based laser weighing even forty tons.)

Constructing a device that could remain workable in space, generally without maintenance, for many years is extraordinarily difficult.[34] The SBL's chemical laser would create movement as it operated—after all, it has many similarities to a small rocket engine in the basic way it generates energy—making it hard to keep a beam on target.[35] In addition to dealing with jitter, the telescopic system would have to point the beam with remarkable accuracy.[36]

These challenges may or may not prove surmountable within two decades. But absent major breakthroughs in materials or rocketry, or both, the costs of building and orbiting a constellation of space-based lasers may prove excessive, even if the concept proves workable. Even if certain new laser concepts, such as the free-electron laser or all-gas-phase iodine laser, were developed in the megawatt range by then, the construction and launch costs of the basic optics alone could still be staggering. Costs for a constellation of two dozen laser weapons were recently estimated at $50 billion or more by the Congressional Budget Office.[37]

Communications Systems

Not all key space-related military technologies are weapons. Indeed, as noted, today the most important are generally those that image Earth or allow rapid, high-data-flow, long-range communications. What are the prospects for further improvements in existing communications and navigation systems that use electromagnetic energy to send signals? What are the prospects for laser communications and other innovative concepts?[38]

Radio communications may be the closest to technological maturity of any of the major categories of technologies considered here. Current limitations on electromagnetic bandwidth and data transmission rates are set by the laws of physics more than by the state of contemporary science and engineering. Indeed, given that future adversaries may well have much greater capacity to disrupt information flows than the likes of Iraq, Serbia, and the Taliban have had in recent years, the U.S. military may have to assume less radio bandwidth in the future rather than more.

Progress will still occur, but perhaps in indirect ways. Data will be more effectively compressed, allowing a given number of photos or a given video stream to be transmitted using several times less bandwidth than is the case today.[39] The priority associated with different types of data transmissions will be more clearly established, allowing the military to quickly forgo luxuries such as frequent teleconferencing in times of war. Continued improvements in the computational power of small computers will allow more on-board processing of raw data on many of the sensor platforms obtaining data. But these types of improvements will do well to slow the growth in the U.S. military's demand for more bandwidth; it is quite unlikely that they will reverse it. Meanwhile, commercial pressures may actually encroach on military bandwidth—there is discussion in the United States of moving the armed forces out of the 1755 to 1850 megahertz zone.[40]

Laser communications systems are another matter, however. They can transfer information at very high data rates, since they operate at high frequencies relative to radio waves. They are often constrained by weather or other atmospheric effects, particularly if they have to communicate with individual ground stations. But for communications among satellites, for communications in good weather at certain wavelengths, or for communications able to make use of distributed ground stations connected by high-speed links, they can easily transmit hundreds of times more information than radio signals.[41] To ensure that ground stations will be available in convenient and cloudless locations for downlinking and then disseminating data, the United States also plans to develop a global information grid (GIG) bandwidth expansion program with a total of ninety sites interlinked by fiber-optic cable.[42]

To be specific, laser communications systems capable of transferring more than 1.0 gigabit per second (Gbps)—more than the entire bandwidth used in the Kosovo or Afghanistan wars—have been under development in recent years. A laser communication link transmitting at 2.5 Gbps was tested at Livermore labs in California in February 2002.[43] It may well be feasible to attain data rates of 10 Gbps or even 30 Gbps. Laser systems also have the advantages of being difficult to intercept or jam, or for that matter even detect, and of using smaller antennas, given their short wavelengths.[44]

Some future systems may route information using lasers in space and then break the information into smaller "packets" to reradiate to Earth with directional antennas using traditional radio waves or microwaves. UAVs above cloud cover could also provide a type of relay. That could make for a truly impressive new type of capability. Such a laser system is now under development in the Pentagon, which is asking for $775 million in 2005 for such a "T-sat" program, though cost growth and technical challenges may delay its arrival until after 2010.[45]

Launch Vehicles, Interceptor Rockets, and Kinetic Energy Systems

A host of space systems rely for their performance on rocketry technology: propulsion systems, aerodynamic technologies, guidance systems, maneuvering capabilities. These are the basic ingredients in launch systems that place objects in orbit, interceptors that would be used to destroy ballistic missiles or satellites through collision or explosion, and a number of more futuristic technology concepts, as well.

Fundamental improvements in the efficiency and cost of space launch systems have been elusive for many years now. Systems continue to improve in some ways, but basic physics and engineering considerations limit progress.[46] Progress in propellants and structural materials for rockets, be they launch vehicles or ICBMs and SLBMs or interceptors, has been limited.[47] Indeed, the theoretical maximum performance of current chemical fuels is being approached. New materials used in the structures of rockets can improve performance at the margin, but major improvements are unlikely with current technology.[48]

It is doubtful that this situation will change much in the coming one to two decades, even in light of the expected infusion of new funds for President Bush's proposed Mars mission. The evolved expendable launch vehicle (EELV) program, the major U.S. effort of late to achieve greater efficiencies and lower costs in space launch operations, will do very well to reduce costs by half.[49] In fact, it seems more likely that it will do well to reduce costs at all; despite the successful launches of the Atlas V and Delta IV in 2002, the financial prospects of Lockheed Martin and Boeing are not being immediately aided by their newest launch vehicles (and Boeing's legal troubles with the government are causing it further problems).[50] These EELV rockets do offer some simplification in numbers of parts and manufacturing processes, but their basic components and efficiencies do not reflect radical improvement.[51] Recent news suggests a large price increase, in fact.[52]

Some innovative concepts may help at the margin for certain applications. For example, there is an interesting idea afoot to launch small satellites (ten kilograms or so) from balloons to allow the use of smaller rockets (the benefit arises from reducing air drag and hence the need for structural strength and weight in the rocket body).[53] On the whole, however, rocket technology is not advancing very fast at present.

But even existing rocket capabilities, when juxtaposed with greater computing power and better sensors, may offer new capabilities. The advent of hit-to-kill technology, beginning in the 1980s (for example, with the short-lived U.S. direct ascent ASAT program) and accelerating significantly in the 1990s, reflects a new accuracy and quickness in sensors, computing, and resulting course adjustment for small "kill vehicles." For example, the midcourse system for missile defense, begun under the Clinton administration, deploys four small "divert thrusters" on its 140 pound exoatmospheric kill vehicle (EKV). It has already struck its target on several occasions, revealing the remarkable quickness and precision of a device that is trying to "hit a bullet with a bullet" (notwithstanding other potential limitations in the system due largely to the likely effects of enemy decoys).[54]

Advances in processing power and miniaturization could also make a concept like brilliant pebbles more feasible than in the past. The idea is to base small interceptors in space for ballistic missile defense, igniting their boosters when necessary to attack a ballistic missile or its warheads. To make such a brilliant pebbles concept work, the interceptors will need to be very fast (to reach ICBMs while they are still burning). Or they will have to possess greater processing power and better sensor technologies—given the low temperatures in space and the fact that warheads would not be identified by their rocket plumes once boost phase was complete.[55] But the concept may prove feasible. Indeed, the Missile Defense Agency is hopeful that a concept for a boost-phase interceptor can be developed within half a dozen years, initially using ground-based rockets but perhaps shortly thereafter for space-based interceptors as well. Those timelines do not seem wildly optimistic.[56] That said, making a single brilliant pebble technically feasible is a far cry from populating low-Earth orbit with enough of them to provide even a limited national missile defense capability. Because such pebbles would always be in motion relative to Earth, and because only a pebble that was near a ballistic missile at the time of launch could destroy it, given the short timelines available for intercept, several dozen pebbles would be needed in orbit for every missile that might need to be destroyed (see table 2-1).[57]

Even more futuristic weapons are being contemplated by defense planners. For example, space-to-Earth kinetic energy attack weapons could also be of interest. The basic science of these types of vehicles is not particularly challenging. However, a dedicated program to create the appropriate types of aerodynamic vehicles would be needed, as would testing. It would be necessary either to develop objects that would fall predictably through the atmosphere without deviating from planned trajectories or burning up, or to develop an aerial vehicle that could fly to its destination once it had been decelerated. But orbiting weapons and later deorbiting them does not offer advantages in speed or cost or technological feasibility, compared, for example, with ballistic missiles.[58] Moreover, hypersonic missiles may ultimately offer a less provocative alternative.[59]

Other new applications of existing technologies can be imagined. To take an extreme example, notice of a large asteroid headed toward Earth could require a dramatic response and crash program, perhaps involving nuclear weapons aboard long-range rockets.[60] But in the shorter term, improving space surveillance is probably the right way to begin to face this unlikely, if extremely worrisome, concern.

Microsatellites and Satellite Swarms

Progress in electronics and computers, as well as improvements in miniaturized boosters, have made possible increasingly small satellites in recent years. These types of devices augur a whole new era in satellite technology. One type of application could be small, stealthy space mines that could position themselves near other countries' satellites, possibly even without being noticed, awaiting commands to detonate and destroy the latter. They could also use microwaves, small lasers, or even paint to disable or destroy certain satellites. Moreover, they could be orbited only as needed, permitting countries to develop ASAT capabilities without having to place weapons in space until they wished to use them.[61]

Most devices known as microsatellites weigh ten to one hundred kilograms; nanosatellites are smaller, weighing one to ten kilograms; picosatellites are even smaller.[62] In recent years, experimental picosatellites—devices weighing less than one kilogram—have been orbited. Two have been put up by the United States; there may be others in space as well, as yet undetected. But it is microsatellites that are becoming prevalent. For example, Germany, China, and the United States have all orbited satellites weighing about seventy kilograms, Brazil has put up a satellite of just over 100 kilograms, and Thailand and Surrey Satellite Technology in the United Kingdom have jointly orbited a device weighing less than fifty kilograms.[63] Advanced microsatellite programs, designed largely for research purposes but also for activities such as communications, are under way in the United States, the United Kingdom, France, Russia, Israel, Canada, and Sweden. Other countries collaborating with private firms based in these locations include China and Thailand, as noted above, as well as South Korea, Portugal, Pakistan, Chile, South Africa, Singapore, Turkey, and Malaysia.[64]

Using microsatellites as ASATs may already be theoretically within near-term reach for a number of countries. The maneuvering capability needed to approach a larger satellite through a co-orbital technique is not sophisticated, especially if there is no time pressure to attack quickly and the microsat can approach the larger satellite gradually. In June of 2000, for example, the University of Surrey launched a five-kilogram nano-satellite built for less than $1 million on a Russian booster (that also carried a Russian navigation satellite and Chinese micro-satellite). The nanosatellite then detached from the other systems and used an on-board propulsion capability to maneuver and photograph the other satellites with which it had been orbited.[65] In early 2003, a thirty-kilogram U.S. microsat maneuvered to rendezvous with the rocket that had earlier boosted it into orbit.[66] Of course, these microsats were already near the satellites they approached, by virtue of sharing a ride on the same booster. But the principle of independent propulsion and maneuvering is being established. Larger maneuvering space mines quite likely are already within the technical reach of a number of countries; smaller ones may be soon.[67]

Space mines could operate in a variety of ways. For example, rather than using explosives, they could carry a high-powered microwave device or the ability to generate a current of strong electricity—or even spray paint. Large high-powered microwave devices could theoretically produce trillions of watts of energy, comparable to the U.S. power grid in output (for a very brief moment). Small devices on microsats would be far less powerful, but if maneuvered to the immediate proximity of a satellite, potentially still quite harmful.

If attacked with such a space mine, be it large or small, the United States might be able to determine its source by inferring which rocket launch had at least roughly positioned the mine near its own satellite. But this detective effort might only happen after the fact, meaning that the United States would probably not be able to prevent the attack. And a patient enemy might be able to launch the microsat into a relatively distant location (at a comparable altitude to its target) and gradually position it properly for intercept, making such detective work even harder. Electronically hardening satellites, as is commonly done for GEO assets because of the intensity of radiation at that altitude, could probably not protect against relatively modest explosive charges. Most countries capable of sending 1,000-kilogram nuclear payloads over intercontinental distances, or of putting relatively large satellite payloads into LEO or MEO, could probably accomplish this task.

The promise of small satellites for a number of applications is real, even if it would be a mistake to assume extremely rapid progress. Certainly, putting small objects into orbit is not a challenge. Providing them with independent propulsion systems capable of substantial maneuvers, as well as adequate guidance packages and sensors to make them useful as ASAT devices, would require work, even for the United States.[68] But it is clearly feasible.

Microsatellites may have benefits to the United States' military uses of space for reconnaissance and communications in addition to their potential use as weapons. For example, large antennas or large mirrors—which are very expensive and are likely to remain so—might be replaced with a number of smaller components acting as an array.[69] They could in theory maintain their positions precisely within the array, using navigation devices and small on-board boosters. (Another approach to reducing use of large, heavy antennas in the future would be to replace structurally dense and solid devices with various types of membranes and tethers that could be sufficient to maintain a solid shape in the low-gravity reaches of space.)[70]

Challenges abound for the idea of satellite swarms, however. Developing the necessary processing power for the signals collected by the individual microsat components, as well as ensuring the synchronization and positional accuracy of the individual components, are quite difficult tasks. This technology is in its infancy and has a long way to go before being shown to be practical, as the Air Force has recently acknowledged in slowing a program, known as TechSat21, designed to investigate this concept.[71] Even if practical, it may not prove cost effective.

Indeed, even much simpler new satellite concepts are proving difficult to realize. For example, the hope of building a Discoverer II constellation of several dozen satellites that would maintain constant coverage of all militarily significant parts of Earth's surface was predicated largely on the belief that satellites could be built for $100 million each. Today, satellites of a similar size cost closer to $1 billion, however, and again, evolution in existing technology seems highly unlikely to produce a tenfold (or even twofold) reduction in costs.[72] The Future Imagery Architecture (FIA) concept may produce its first launch only in 2007 or 2008, rather than 2006, and more than $3 billion had to be added to the program's expected price tag in 2003.[73] The space-based infrared system-high and SBIRS-low programs have each more than doubled in cost since 1996 and continue to face significant technical challenges with their infrared sensors, communications systems, and weight. The former is designed principally for warning of missile launch, the latter for tracking missiles and warheads in space for ballistic missile defense. First launch has slipped from 2002 to 2007 for SBIRS-high, and for the SSTS system (another, newer term for SBIRS-low), from 2004 to 2007. If these problems are harbingers of the difficulties of developing much smaller satellites, their day is probably still quite far off.[74]

Conclusion

The basic physics of the military use of space is challenging, and change is not happening quickly in most major technology sectors. Rocket boosters are only slightly improved from those of two or three decades ago, space-based lasers appear a distant and very expensive prospect, space-to-Earth weapons have limited intrinsic appeal, even if they probably could be built. But some areas of technology, such as high-energy lasers and microsatellites, are developing quickly enough that they could substantially change the basic backdrop for making military space policy in the coming decade or so. And existing technologies are spreading to many more users, as well.

CHAPTER FOUR

A FUTURE TAIWAN STRAIT CONFLICT

Preceding chapters have summarized the state of debate over the future military uses of space, the use of space today for civilian and military purposes, and trends in underlying technologies that are likely to change the policy backdrop in coming years. To bring these various considerations together prior to considering future policy options for the United States in chapter 6, it is helpful to analyze a specific scenario in which military space assets could play a critical role. This chapter considers one such possibility: a war between the United States and China over Taiwan, set in the period 2010 to 2015.

Other countries could also pose concerns to the United States at that time, of course. For example, Iran is reportedly already considering its own reconnaissance satellite.[1] But a China scenario is perhaps the most challenging to the United States, given that country's resources, the extent of its interest in space, and its awareness of American capabilities in space. Consideration of this scenario does not suggest it is likely; indeed, there is good reason to think China and the United States will avoid military competition in the years ahead and gradually improve their relations. But a prudent military planner must recognize the possibility of crisis over Taiwan, nonetheless.

The following issues are of greatest interest. Would China have a meaningful ability to impede U.S. warfighting goals using antisatellite weapons? Might the United States need ASATs to prevent China from tracking valuable large assets, such as aircraft carriers? And how important would all these issues be in the context of the overall military balance?

This chapter's central answer to these questions is sobering, yet not alarmist. China may well have the ability to impede the United States' use of satellites within a decade or so. It could likely jam some commercial communications satellites on which the U.S. armed forces increasingly depend in wartime. It could almost surely threaten low-Earth orbit imaging satellites with at least one of several possible types of weapons—nuclear-tipped missiles (a real concern, since they could be used in space without killing many people, if any), lasers, and possibly microsatellites. It may be able to threaten a small number of key military satellites in geosynchronous orbit as well—such as American electronic eavesdropping devices—with microsatellites. In addition, China could well be capable of finding and targeting large American military assets such as aircraft carriers, at least occasionally, using its own space assets and command/control networks. It could do so perhaps via the periodic overflight of an imaging satellite, which might be able to keep a carrier in its field of view long enough to get coordinates to a cruise missile launched from a submarine or aircraft. Or it might use a constellation of radar reconnaissance satellites more akin to what the Soviet Union used to employ.[2] (It is not trivial to build such constellations and make them work—malfunctions reportedly prevented the Soviets from monitoring the 1982 Falklands War, for example, despite considerable investment—but the technology is well established and understood.)[3]

At the same time, the United States need not be immediately and equally concerned about all of these potential threats. Neither the China-Taiwan scenario described below nor any other justifies the use of the term "space Pearl Harbor." The threats that are most likely to become imminent can be countered by improving and hardening certain U.S. satellite capabilities (as discussed further in chapter 6) and by ensuring adequate airborne breathing backups to satellites. The ability to threaten Chinese satellites, should any need arise, can be ensured by continuing U.S. programs for jamming PRC communications links and, as a hedge, by proceeding as planned with missile defense programs that have latent ASAT capability. That would allow for rapid adaptation of missile defense systems to antisatellite weapons, should that become necessary. However, it is not necessary at present, and sufficient warning can be expected to allow for policy response, should that become imperative.

China's Current Military Capabilities

China's military has many weaknesses, and the threat it poses can be exaggerated. That said, China will improve its military and could well be a future security challenge to the United States. Moreover, any conflicts that pitted China against the United States would probably occur close to the PRC's shores, giving China a number of advantages. Finally, the Taiwan issue remains of great concern to China and is capable of again deteriorating into crisis, given views in Taipei and Beijing.

China has the world's largest military, by far. It roughly ties with Russia, Japan, the United Kingdom, and France for the claim to second greatest level of military spending in the world.[4] That conclusion is based on the most widely accepted estimates of its actual expenditures (as opposed to its misleadingly low official figures). Taiwan has a much smaller, and considerably less expensive, military—but is still about tenth in the world in total defense spending, and its reserve forces are actually larger than those of China. Taiwan's troops are generally better educated, trained, and equipped than China's, even though they fall short by some standards themselves.

China's military has traditionally focused on internal and border security much more than on foreign operations. Of China's nearly 2 million ground troops, only about 20 percent are considered by the Pentagon to be mobile, even within mainland China itself. Considerably fewer could deploy abroad, given their dearth of logistics assets (for example, trucks, construction and engineering equipment, mobile depots and hospitals, and fuel-storage infrastructure). Even though Taiwan is only about 100 miles away from mainland China, the fact that it is separated by water further constrains the PRC's ability to project military power there. Few PRC troops could deploy over water, given China's very limited military airlift and sealift capacity. Its seventy or so amphibious ships could move about 10,000 to 15,000 troops with their equipment, including perhaps 400 armored vehicles; airlift could move another 6,000 troops, or perhaps somewhat more, counting the possibility of helicopter transport as well.[5]

These shortfalls in transport and logistics would be magnified by China's other military weaknesses. Its training, and the overall caliber of its armed forces, leave much to be desired. Although Chinese military personnel are generally competent at basic infantry skills, the armed forces do not tend to attract China's best, nepotism is prevalent, party loyalty is of paramount importance, most soldiers are semiliterate peasants serving short tours of duty, and a strong professional noncommissioned officer corps is lacking. Combined-arms training, while somewhat enhanced of late for elite rapid-reaction forces, is infrequent, and joint service training remains rare. Specialized assets, such as aerial refueling, electronic jamming, and command aircraft, are in short supply and of mediocre quality. The Chinese military's aspirations to conduct "local wars under high-technology conditions" are far from being realized, and its capabilities for taking advantage of the so-called revolution in military affairs, while much ballyhooed, are in all likelihood quite limited.[6]

These overall realities are unlikely to change soon. China's indigenous defense industry is of mediocre caliber. Much of the country's defense budget must be devoted to paying, training, and supplying its large numbers of troops. For such reasons, and given China's limited defense resources—especially when measured against such a large military—the Defense Intelligence Agency estimates that only 10 percent of China's armed forces will have "late–cold war" equivalent hardware even by 2010. Its large attack submarine force includes only about a dozen submarines that could be viewed as relatively modern— and half of those, the nuclear-powered Han vessels, are rather noisy and unreliable. The People's Liberation Army Air Force (PLAAF) is projected to add only twenty to thirty top-notch fighter aircraft to its forces annually in the years ahead. Moreover, there are doubts about China's ability to maintain and effectively operate whatever modest number of advanced fighter jets it is able to acquire.[7] These facts cast doubt on China's ability to establish air superiority in a hypothetical war against Taiwan even in five or ten years or to compete favorably with Taiwanese ground forces, should China ever manage to establish a toehold on a Taiwanese coast.

But China could be a threat to Taiwan for other reasons. It could try to attack Taiwan with missiles, for example. A missile strike seems rather plausible, given China's ongoing buildup of missiles near Taiwan and its ability to adjust rather precisely the amount of force used with such weapons. However, assuming that such missiles would be armed only with conventional warheads, such an attack may not be particularly effective in forcing Taiwan to capitulate. In fact, it might leave China with no obvious next escalatory step—even as it probably unified the Taiwanese people and increased their desire to resist.

A Chinese naval blockade of Taiwan could be more effective, threatening the sea routes on which the island's economy fundamentally depends. Even if the blockade is "leaky," it could convince many shippers not to risk the journey. Breaking it would require a forcible convoy operation, as well as a de-mining operation likely to remain beyond the capacity of Taiwan's navy even in 2010 or 2015 and hence necessitating U.S. involvement.[8]

On balance, the United States could handle the types of threats that China might most plausibly pose in coming years. It could not stop all missiles from striking Taiwan, of course, but it would have a good chance of breaking a naval blockade. Such a scenario would play out in the waters and airways of the Western Pacific, where the United States and its allies are strongest, rather than on the landmass of Asia. The United States would still have difficulty handling the challenge and could need a substantial fraction of its force structure (especially its naval and air capabilities) to prevail.[9] Assuming no escalation to nuclear warfare or to acts of state-sponsored terrorism, it could probably win any battles with relatively limited loss of life and relatively little uncertainty about the outcome. It could keep most of its key military assets in the waters east of Taiwan. China would have a very difficult time getting airplanes or submarines (not to mention surface ships) into that region without having those assets preemptively destroyed by the United States in time of war. And if it did not get such military platforms into that region in considerable numbers, China could not find or attack most U.S. assets—except, occasionally, any smaller vessels that were escorting Taiwanese ships into port.

The Role of Space in Future War

But what about the situation in 2010 or 2015? As the above analyses and projections suggest, China is quite unlikely to have a first-rate military by that point. However, China may not need to approach U.S. capabilities to have a plausible chance—at least in its own leaders' eyes—of prevailing in war.

It is doubtful that trends in space capabilities or any other aspect of defense modernization will radically alter the basic military balance in the next decade or so. The size and caliber of the U.S. military is sufficient that, even if China were able to close the technological gap and have the potential to cause substantial losses to the United States in a war over Taiwan, the American armed forces would still surely prevail. The United States could lose a carrier or two and still maintain overwhelming military superiority in the region.

But there is nonetheless a worrying feature of such a gradual shift in the military balance. Given trends in military reconnaissance, information processing, and precision-strike technologies, large assets such as aircraft carriers and land bases, on which the United States depends, are likely to be increasingly vulnerable to attack in the years ahead. Land bases can to an extent be protected, hardened, and made more numerous and redundant, but ships are a different matter. How fast, and whether, China can exploit these trends remains unclear, as noted above. But the trends are real nonetheless.[10] As a recent example, China reportedly tested an antiship cruise missile that proved to have twice the range originally expected by U.S. intelligence—155 miles.[11] And its space assets are surely growing in scope. Even if it does not have an extensive imaging satellite network in a decade or so, it may be able to orbit one or two reconnaissance satellites that could occasionally detect large ships near Taiwan. That might be good enough. If China could find major U.S. naval assets with satellites, it would only need to sneak a single airplane or ship or submarine into the region east of Taiwan to have a good chance of sinking a ship.

Knowing the U.S. reluctance to risk casualties in combat, China might convince itself that its plausible ability to kill many hundreds or even thousands of U.S. military personnel in a single attack would deter the United States from entering into the war in the first place. Such a perception by China might well be wrong (just as Argentina was wrong to think in 1982, in a somewhat analogous situation, that it could deter Britain from deciding to take back the Falkland Islands); but it could still be quite dangerous, given the resulting risks of deterrence failure and war.

China is certainly taking steps to improve its capabilities in space operations. According to the Pentagon's latest assessment, "Exploitation of space and acquisition of related technologies remain high priorities in Beijing. China is placing major emphasis on improving space-based reconnaissance and surveillance.

... China is cooperating with a number of countries, including Russia, Ukraine, Brazil, Great Britain, France, Germany, and Italy, in order to advance its objectives in space." China will also surely focus on trying to neutralize U.S. space assets in any future such conflict; no prudent military planner could do anything else. According to the Pentagon,

> Publicly, China opposes the militarization of space, and seeks to prevent or slow the development of anti-satellite (ASAT) systems and space-based ballistic missile defenses. Privately, however, China's leaders probably view ASATs—and offensive counterspace systems, in general— as well as space-based missile defenses as inevitabilities. ... Given China's current level of interest in laser technology, Beijing probably could develop a weapon that could destroy satellites in the future. .... A Hong Kong newspaper article in January 2001 reported that China had developed and tested an ASAT system described as a "parasitic microsatellite."[12]

Exactly how many U.S. satellites, and of what type, China might be able to damage or destroy is hard to predict. But it seems likely that low-altitude satellites as well as higher-altitude commercial communications satellites would be vulnerable. Low-altitude imaging satellites are vulnerable to direct attack by nuclear-armed missiles, at a minimum, by high-energy lasers on the ground, and quite possibly by rapidly orbited or predeployed microsatellites as well. They are sufficiently hardened that they would have to be attacked one by one to ensure their rapid elimination. And they are sufficiently capable of transmitting signals through or around jamming that China probably could not stop their effective operation in that way. But they are few enough in number, and sufficiently valuable, that China might well find the means to go after each one.

For higher-altitude military satellite constellations, including GPS, military communications, and electronic intelligence systems, China's task would be much harder. Such constellations often have greater numbers of satellites than do low-altitude imagery systems. They are probably out of range of most plausible laser weapons, as well as ballistic missiles carrying nuclear weapons. They might, however, be reached by microsatellites deployed as hunter-killer weapons, particularly if those microsats had been predeployed (a few might be orbited quickly just before a war, but launch constraints could limit their number, since microsats headed to different orbits would probably require different boosters).

Finally, high-altitude commercial communications satellites are quite likely to be vulnerable. Their transmissions to Earth might well be interrupted for a critical period of hours or days by jamming or a nuclear burst in the atmosphere. For example, disruption of UHF radio signals due to a nuclear burst can last for many hours over a ground area of hundreds or even thousands of kilometers per dimension. Unhardened satellites might be damaged by a large nuclear weapon even at distances of 20,000 to 30,000 kilometers.[13] They might even be vulnerable to laser blinding.

So it appears that China will remain quite far behind the United States in military capability, relatively rudimentary in its space capabilities, and lacking in sophisticated electronic warfare techniques and similar means of disrupting command and communications. But it could hamper some satellite operations. And it could have an "asymmetric capability" to find, target, and attack U.S. Navy ships (not to mention commercial ships trying to survive the postulated blockade of Taiwan).

Some might argue that the above analysis overstates the potential role of satellites. For example, even if China would have a hard time getting aircraft close enough to track U.S. ships, given American air supremacy, it might have other means. For example, it may be able to use a sea-based acoustic network. Most probably such a system would be deployed on the seabed, as with the U.S. SOSUS array.[14] On that logic, China may have so many options and capabilities that it need not depend on any one type, such as space assets.

Or China may not be able to make good use of any improvements it can achieve in its satellite capabilities. To use a reconnaissance-strike complex to attack a U.S. carrier, one needs not only periodic localization of the carrier, but real-time tracking and dissemination of that information to a missile that is capable of reaching the carrier and defeating its defenses. The reconnaissance-strike complex must also be resilient in the face of enemy action. The PRC is not close to having such a capability, either its constituent parts or as part of an integrated real-time network.

But the case for concern in general, and for special concern about Chinese satellite capabilities, is still rather strong. If China does improve its satellite capabilities for imaging and communications, the United States could be quite hard-pressed to defeat them without ASAT capabilities. Destroying ground stations could require deep inland strikes—and may not work if China builds mobile stations. The sheer size of the PRC also makes it difficult to jam downlinks; the United States cannot flood all of China continuously with high-energy radio waves (though it may be able to jam links to antiship cruise missiles already in flight, if it can detect them, it would be imprudent to count on this defense alone). Jamming uplinks may be difficult as well, if China anticipates the possibility and develops good encryption technology or a satellite mode of operations in which incoming signals are ignored for certain periods of time. Jamming any PRC radar-imaging satellites may work better, since such satellites must transmit and receive signals continuously to function. But that method would work only if China relied on radar, as opposed to optical, systems.

In regard to the argument that China could use SOSUS arrays or other such capabilities to target U.S. carriers, making satellites superfluous, it should be noted that the United States has potential means for countering any such efforts. To deploy a fixed sonar array in the vast waters east of Taiwan where U.S. ships would operate in wartime, China would need to predeploy sensors in a region many hundreds of kilometers on a lateral dimension at least. This could be technically quite difficult in such deep waters. Although the United States has laid sonar sensors in waters more than 10,000 feet deep, the procedure is usually carried out remotely from a ship or by a special submarine, and hence is increasingly difficult as depth increases.[15] In addition, the United States would have a very good chance of recognizing what China was doing. Even though peacetime protocols would prohibit preemptive attacks, the United States could be expected to know where many of China's underwater assets had been deployed, allowing attacks of one kind or another in wartime. The United States is devoting considerable assets to intelligence operations in the region already, for example, with its attack submarine force.[16] It would similarly have a good chance of detecting and destroying Chinese airborne platforms, including even small UAVs, used for reconnaissance purposes.

On balance, growing Chinese satellite capabilities for targeting and communications could be an important ingredient in what Beijing might take (or mistake) for a war-winning capability in the future. China would not need to think it had matched the U.S. armed forces in most military categories, only that it had an asymmetric ability to pose greater risks to the United States than Washington might consider acceptable in the event of a future Taiwan Strait crisis.

China might also have means to attack U.S. space assets, particularly lower-flying reconnaissance satellites, by 2010 or 2015. It is not entirely out of the question that China might use a nuclear weapon to do so, knowing that such a strike might greatly weaken U.S. military capabilities without killing many, if any, Americans. China attaches enough political importance to holding onto Taiwan that it might well prove quite willing to run some risk of escalation in order to do so—especially if its leaders thought they had deduced a clever way to escalate without inviting massive retaliation. Whether it could disrupt or destroy most satellites is unclear; whether it could reach large numbers of GPS and communications assets in medium-Earth orbit and geosynchronous orbit is doubtful. But for this and other reasons it is also doubtful that the United States could operate its space assets with impunity, or count on completely dominating military space operations, in such a scenario.

Conclusion

The United States is not in danger of falling behind China, Iran, or any other country in military capability in the coming years and decades. And its own capabilities will probably grow, in absolute terms, faster than those of any other country.

But its relative position could still suffer in a number of military spheres, including space-related activities. Its satellites will be less dependable in conflict than they are today, or have been in recent years; other countries may also mimic the U.S. ability to use satellites and accompanying ground assets for targeting and real-time attack missions. The trends are not so unfavorable or so rapid as to require urgent remedial action; indeed, the United States has military and political reasons to show restraint in most areas of space weaponry. But passive defensive measures should be expanded and some potential offensive capabilities investigated so as to retain the option of weaponizing them in the future, if necessary. These topics are the subjects of chapter 6.

CHAPTER FIVE

ARMS CONTROL
IN SPACE

Should the United States agree to restraints on the future military uses of outer space, and in particular the weaponization of outer space? That is, should it sign treaties prohibiting the testing, deployment, or use of weapons in space or of Earth-based weapons that might be used against objects in space? And in cases where treaties do not make sense, could less formal approaches be useful?

Any formal treaties would probably have to be multilateral. It makes little sense to consider bilateral treaties because it is unclear which country would be the other party. At this point, any space treaty worth the effort to negotiate would probably have to include as many other space-faring countries as possible, ranging from Russia to European powers to China to India to Japan. The fact that any accords would be multilateral does not mean that they should be negotiated under the auspices of the United Nations. There is a strong and perhaps ideological pro–arms control bias in the UN forums where many space arms control discussions have occurred to date. In addition, it appears that some countries may be using those forums to embarrass the United States rather than genuinely to pursue long-term accords to promote international stability. The United Nations might ultimately be involved to bless any such treaty, but it might be best to negotiate it in some other context. Russia and China have recently expressed more flexibility on the format of any possible talks, no longer insisting on formal negotiations leading to a treaty at the Conference on Disarmament.[1] But one must still decide if such treaties are worth negotiating in the first place.

The question of space arms control tends to elicit immediate reactions from the left and the right—the former for, the latter against most restraints. However, the subject is important enough, and complex enough, to merit more than an instinctive reaction. Treaty skeptics may at least want to consider bans on certain kinds of space-based weapons that the United States would be unlikely ever to use, such as antisatellite weapons producing large amounts of debris. They also may wish to ask themselves if unilateral, informal, and temporary restraints on U.S. space activities could advance American interests by helping to preserve the space status quo—in which weaponization is absent and the United States benefits from unrivaled targeting and communications assets in orbit. Treaty proponents may not want to put themselves in the position of supporting accords that are inherently unverifiable or easy to circumvent in other ways. Many other aspects of the possible weaponization of space merit careful attention as well.

This chapter briefly categorizes, summarizes, and assesses the various main options in space arms control that the United States and the international community might consider in the future. Only a couple of them merit strong American support at this point, in my judgment. But those couple may be important and should be seriously pursued. Specifically, the case is persuasive for an arms control accord banning debris-causing antisatellite tests in space. A more informal code of conduct that would discourage testing of any ASAT weapon against any satellite for the foreseeable future also makes sense, though this accord should not be formal or permanent, since circumstances may someday change. The United States should also consider revising its space doctrine to forswear near-term development of dedicated ASAT systems, even short of testing. And a debate on whether to permanently ban space-to-Earth weapons should also begin; this is a sufficiently futuristic matter that urgent attention is not required, however.

Ideas for space arms control may be grouped into three broad categories. First are outright prohibitions of indefinite duration. The existing Outer Space Treaty's bans on nuclear weapons in space and military colonies in space are examples of this type of accord. Prohibitions can be broad. They can also be specific—such as bans on the testing or deployment of antisatellite weapons above certain altitudes or bans on destructive ASATs that would produce long-lasting space debris. This category might also include broad bans of temporary duration. Second are confidence-building measures, such as advance notifications of space launches and keep-out zones around deployed satellites. Third are informal understandings, worked out in talks or more likely established through the unilateral but mutual actions of major state powers. An example might be a decision to forgo the testing and deployment—and perhaps even the development—of dedicated ASATs for the foreseeable future (table 5-1 sets out means of verification to monitor such an ASAT test ban).

Table 5-1. Current and Potential National Technical Means of Verification for Monitoring a Comprehensive ASAT Test Ban

ASAT type	Current means of verification				Potential means of verification		
	Reconnaissance satellites	Signals-intelligence systems	Ground-based radar	Ground-based electro-optical systems	Space-based long wavelength infrared sensors	Space-based multi-spectral imagers	Gamma ray spectrometers
Ground-based kinetic energy weapon	Launch site activity	Telemetry; command and control	Maneuvers; impact debris	Maneuvers; impact debris	Maneuvers; impact debris	n.a.	n.a.
Ground-based high-energy laser	Test site activity	n.a.	Debris	Debris	Thermal radiation	Thermal radiation	n.a.
Conventional space mine	n.a.	Telemetry; command and control	Maneuvers; impact debris	Maneuvers; impact debris	Maneuvers; impact debris	n.a.	n.a.
Nuclear space mine	n.a.	Telemetry; command and control	Maneuvers	Maneuvers	Maneuvers	Gamma radiation	Gamma radiation
Space-based high-energy laser	Functionally related characteristics	Telemetry; command and control	Functionally related characteristics	Functionally related characteristics	Thermal radiation	Thermal radiation	n.a.
Space-based neutral particle beam	Functionally related characteristics	Telemetry; command and control	Functionally related characteristics	Functionally related characteristics	n.a.	n.a.	Gamma radiation from target

Source: Paul B. Stares, *Space and National Security* (Brookings, 1987), p. 164.

Prohibitions

One type of arms control accord on activities in space would be quite comprehensive—no testing, production, or deployment of ASATs of any kind, whether based in space or on the ground, at any time; no Earth-attack weapons in space, ever; with formal treaties of permanent duration codifying these prohibitions. These ideas are in line with proposals made by the Chinese and Russian delegations to the UN Conference on Disarmament in Geneva. They also are made by arms control proponents, such as Ambassador Jonathan Dean, who believe that space should be a sanctuary from weaponization and that the Outer Space Treaty already strongly suggests as much.[2]

There are three main problems with such ideas, however. To begin with, it is difficult to be sure that other countries' satellite payloads are not ASATs. This is especially true in regard to microsatellites, which are hard to track. Some have proposed inspections of all payloads going into orbit, but this would not prevent "breakout," in which a country on the verge of war would simply refuse to abide by these provisions. Since micro-sats can be tested for maneuverability without making them look like ASATs, it will be difficult to preclude this scenario. A similar problem arises with the idea of banning certain types of experimentation, such as all outdoor experiments or flight test-ing.[3] A laser can be tested for beam strength and pointing accu-racy as a ballistic missile defense device, without being identi-fied as an ASAT. A microsat can be tested for maneuverability as a scientific probe, even if its real purpose is different, since maneuvering microsats capable of colliding with other satellites may have no visible features indicating this intent. Bans on out-door testing of declared ASAT devices would do little to impede development of such capabilities.

Second, in a broader sense it is not possible to prevent certain types of weapons designed for ballistic missile defense from being used as ASATs. This can be considered a problem of verification. But in fact, the issue is less of verification, per se, than of knowing the intent of the country building these systems—and ensuring that that intent never changes. This is unrealistic. In other words, some systems designed for missile defense have inherent ASAT capabilities and will retain them, due to the laws of physics, regardless of what arms control prohibitions are developed, and countries possessing these missile defenses will recognize these latent capabilities.[4] For example, the American midcourse missile defense system and the airborne laser would both clearly have inherent capabilities against low-Earth orbit satellites if given good information on a satellite's location and perhaps some software modifications. The United States could declare for the time being that it will not link these missile defense systems to space surveillance networks or give them the necessary communications and software capabilities to accept such data. But such restraints, while currently worthwhile as informal, nonbinding measures (see below), cannot be easily verified and can be easily reversed. Thus no robust, long-term formal treaty regime should be based upon them. Indeed, the problem goes beyond missile defense systems. Even the space shuttle, with its ability to maneuver and approach satellites in LEO, has inherent ASAT potential. So do any country's nuclear weapons deployed atop ballistic missiles. Explicit testing in ASAT modes can be prohibited, but could also have limited meaning.

Third, it is not clear that the United States will always benefit militarily from an ASAT ban. The scenario considered in chapter 4 involving a war in the Taiwan Strait is a good example of how, someday, the United States could be put at serious risk by another country's satellites. This day is not near, and there are many possible ways to deal with the worry in the near term other than by developing destructive ASATs. But over time, the need for such a weapon cannot be ruled out.

Finally, consider the longer-term question of Earth-attack weapons based in space. Most such weapons would probably require considerable testing, given the realities and difficulties of atmospheric reentry. That means that prohibitions might well be verifiable. Furthermore, prohibitions on such weapons may cost the United States little, since it will retain other possible recourses to delivering weapons quickly over long distances. The most powerful argument against banning ground-attack weapons in space is that they are a long-term prospect, the need for which cannot be easily assessed now. But the United States can probably make do without them, or find alternatives. So, a ban may someday make sense but is not an immediate priority.

A number of specific ASAT prohibitions, fairly narrowly construed, are worth considering as well. They could be carefully tailored so as not to preclude development of various capabilities in the future. But they could also help reassure other countries about U.S. intentions at a time of still-unsettled great power relations, and help protect space from excessive debris or other hazards to safe use over the longer term. Such measures could include

—temporary prohibitions, possibly renewable, on the development, testing, and deployment of ASATs or Earth-attack weapons,

—bans on testing or deployment of ASATs above certain altitudes,

—bans on debris-producing ASATs, and

—prohibitions against first use of ASATs and space weapons.

Temporary formal treaty prohibitions would be no more verifiable than permanent bans. But they could make sense in certain situations.

There are downsides to signing accords from which one might
very well withdraw, of course. If and when the United States
could no longer support the prohibitions, it would likely suffer in
the court of international public opinion by its unwillingness to
extend the accord—even if it was specifically designed as non-
permanent. The United States' experience with the ABM Treaty
suggests that the damage from such decisions can be limited.
President Bush was able to withdraw from the ABM Treaty without worsening U.S.-Russian ties, given that there were sound
strategic arguments that the case for missile defense had changed
fundamentally and strengthened greatly in the thirty years since
the treaty had been signed. That said, it is hard to withdraw from
treaties, suggesting that, on balance, the United States should not
sign most accords that it already knows to be temporary.

Bans on debris-producing ASATs do make sense, and could
well be codified by binding international treaty of indefinite
duration. Destructive testing of weapons such as the Clinton
midcourse missile defense system or other hit-to-kill or explosive devices against objects in satellite orbital zones would not
only increase the risk of an ASAT competition. It would also
create debris in LEO regions that would remain in orbit indefinitely (unless the testing occurred in what are effectively the
higher parts of Earth's atmosphere, where air resistance would
ultimately bring down debris, and few, if any, satellites fly in
any case). The U.S. military worries about this effect of debris-
producing weapons. Tests of the midcourse system to date have
occurred at altitudes of roughly 140 miles, producing debris
that de-orbits within roughly twenty minutes, but future tests
will be higher. A ceiling of perhaps 300 miles might be placed

on such tests, and a ban imposed on using targets that are in orbit.

Another possible type of treaty regime might ban testing and deployment of all types of ASATs, even those that do not produce debris, above a certain altitude. The natural cutoff region might be around 1,000 kilometers above Earth's surface. This is roughly the ceiling for the flight trajectory of ballistic missile warheads. It is also roughly the distance at which laser beams like the ABL begin to become rather weak. By creating such a ceiling, satellites located at heights of a few thousand kilometers should enjoy considerable protection.[5] This approach suffers from most of the same problems as complete ASAT prohibitions. Its stipulations are largely unverifiable, given the potential of microsats, in particular, to be developed and tested under false pretenses. Its terms may not even be strategically advantageous for the United States, over time.

A ban on the actual use, and particularly the first use, of ASAT systems could be considered as well. The downside of this approach is that in a war in which satellites are being used for tactical warfighting purposes, it is not clear why they should be given sanctuary—and dubious that they would be. On balance, this idea does not seem worthwhile. The ban on debris-causing activities is the main possible prohibition that seems sound today.

Confidence-Building Measures

Another category of arms accords includes those that do not limit the weapons capabilities of states, but instead seek to establish mechanisms for the use of their assets. The goals would be to reduce tension, improve communications, calm nerves, and build safety mechanisms into the military use of outer space. This approach would build on some of the measures the nuclear superpowers took to reduce the potential for unintentional nuclear confrontation during the cold war, including the 1972 Incidents at Sea accord and agreements to set up communications hotlines.[6] Here the stakes might not be so great, but they could still be great enough to justify some straightforward measures and rules of the road.

One such idea is that of keep-out zones around deployed satellites. The concept here is that there is no reason for a satellite to approach within a few tens of kilometers, or in some orbits even within hundreds of kilometers, of another satellite. Any close approach can be assumed to have hostile purposes and thus ruled out as an acceptable action.

What real strategic purpose would be served by such zones? Unless satellites were given self-defense capabilities—tough to distinguish from those of offensive ASATs—they could not be enforced. And any country wishing to develop a close-approach capability for the purposes of ultimately launching a large-scale ASAT surprise attack could do so despite the existence of keep-out zones, by testing against its own space assets or even against empty points in space.

That said, keep-out zones may still make sense, even though they would not constitute a substantial limitation on military capabilities. Creating such zones would add another step that any state undertaking an attack would have to build into its plan. ASATs could not easily be predeployed near other satellites without arousing suspicion (especially if the United States and other countries deployed satellites with sensors capable of monitoring their neighborhoods). Any state violating the keep-out zones would alert the targeted country to its likely intentions; conversely, respecting the zones would constitute a form of restraint that could calm nerves to some modest but perhaps worthwhile degree. On balance, this idea is sound, though not worth a great deal of top-level time to negotiate.

What of advance notice of space launches? Again, this type of accord, such as that reached between the United States and Russia during the Clinton administration, would not prevent a country from breaking out of the accord suddenly. It would not constitute a meaningful constraint on capabilities. But as long as it was observed, countries would have additional reassurance that others were playing by the rules. They would also have time to prepare to observe the deployment of satellites from any launch, allowing slightly greater confidence that ASATs were not being deployed. As a peacetime rule of the road, at least, it makes sense.

Some have also suggested allowing international monitoring of space payloads prior to their launch.[7] This idea seems more questionable, though, since satellites could effectively function as ASATs without carrying payloads that were obviously offensive.

On balance, several of these confidence-building measures seem marginally useful. They will not prevent the United States from retaining its hedges against some future need for ASATs, whether in the form of dual-purpose ballistic missile defense programs or even a dedicated antisatellite system. They will not prevent China or another country from quietly building inherent ASAT capability. But they will add an extra step or two that other countries choosing to weaponize space would need to undertake before threatening American interests.

Informal Unilateral Restraints

A final category of arms control measures would not involve arms control at all—in the formal sense of signed treaties and binding commitments—but unofficial and unilateral restraints. The idea would not be for the United States to tie one hand behind its back while other countries were free to pursue space weapons. Rather, the hope would be to encourage countries to show mutual restraint by setting a precedent and a tone through U.S. action. To the extent some countries did not show restraint, the policy could then be reconsidered.

This approach has a number of precedents in international affairs. For example, countries showed restraint by not using chemical weapons in World War II, even though there was no formal agreement to avoid these terrible agents. More recently, the United States reduced the alert levels of some nuclear forces and took tactical nuclear weapons off naval vessels in the first Bush administration, in part to encourage similar Soviet actions, which then followed.[8] And even in the absence of ratification of the nuclear test ban treaty, almost all countries have continued to respect a moratorium on testing.

Informal restraint can work more quickly than formal arms control. It can also preserve flexibility for the future, should circumstances change. If the United States would have ample time to change its policy in the event other countries failed to cooperate, without harm to its security interests in the interim, there is much to be said for this approach.

Since the United States is not at present building or deploying space weapons, a policy of informal restraint could easily apply to research and development and testing activities. As one example, if a treaty to prevent the creation of space debris through testing of any ASAT could not be quickly negotiated, the United States could make a pledge unilaterally to accomplish this goal.[9] The flexibility associated with such a unilateral pledge might permit the United States to go further and also pledge not to create any ASAT that would ever create such debris, given that even if it needs a future ASAT it would have alternative technological options. (On a related but nonweapons matter, it may also be worthwhile to consider requiring commercial satellite builders to de-orbit old LEO satellites and adopt other debris-mitigation measures as a condition for gaining licenses to put objects into space.)[10]

The United States might also consider making a clear statement that it has no dedicated ASAT programs and no intention of initiating development or deployment of any. It could also declare that it would not test any systems, including high-powered lasers, microsatellites, and ballistic missile defenses, in an ASAT mode. The latter approach would have the greatest chance of eliciting verifiable reciprocation by other countries.

The downside to making such statements is that if and when U.S. policy requirements changed, the statements would have to be repudiated—raising alarm bells abroad and risking greater diplomatic problems than if the United States had never held itself to informal restraints. The advantage is that they might buy the United States some time, allowing it to stigmatize space weapons that it has no strategic interest in developing or seeing developed anytime soon. All of these options are considered in the final chapter of the book.

CHAPTER SIX

PRESERVING U.S. DOMINANCE WHILE SLOWING THE WEAPONIZATION OF SPACE

As established in the preceding chapters, the basic backdrop for devising future U.S. space policy is roughly as follows. First, the United States increasingly uses space for military purposes, particularly for tactical warfighting. It will surely continue to increase its dependency on reconnaissance, targeting, and communications satellites for such activities. Second, although the United States in particular, and certain other countries to a lesser extent, have militarized space in such ways, they have not yet weaponized space. That is, they have not placed weapons in orbit or developed weapons designed to attack satellites.

Third, the ability of the United States and other countries to rely on space systems cannot be ensured indefinitely. Already, the nuclear powers have ballistic missile forces that constitute latent ASAT capabilities. The United States, in particular, also is pursuing several ballistic missile defense programs; other countries may soon have similar, if less technically advanced, capacity. For U.S. armed forces, of particular concern are inherent vulnerabilities in low-altitude imaging satellites as well as commercial communications systems that a moderately capable adversary might eventually be able to exploit using microsatellites, lasers, or even nuclear weapons. Fourth, other countries will gradually gain a greater capability to use space for offensive military purposes. In particular, they are likely to gain the capacity to find and target large mobile assets such as ships and major ground force formations—if not continuously, at least sporadically.

Fifth and finally, more futuristic space capabilities, such as space-to-Earth weaponry or large constellations of space-based lasers for ballistic missile defense, are likely to remain futuristic. But certain exotic concepts such as "brilliant pebbles" space-based ballistic missile defense rockets may be feasible within a decade or so—though their deployment in the numbers needed for missile defense (even against a small threat) will likely remain extremely expensive and inadvisable on budgetary as well as strategic grounds.

Basic technological and strategic realities argue for a moderate and flexible U.S. military space policy. They argue against two extreme positions that have been espoused by prominent U.S. policymakers in recent years. The report of the Commission on Outer Space, which warned of a possible space "Pearl Harbor" and implied that the United States needed to rapidly take many steps—including offensive ones—to address such a purportedly imminent risk, was alarmist. Most U.S. satellites are not vulnerable to attack today and will probably not be in the years ahead. Indeed, the director of the Defense Intelligence Agency testified in early 2003 that even antisatellite methods such as effective jamming, as well as kinetic and directed-energy weapons, will not be easily and widely available in the next ten years.[1] Thereafter, many threats may be handled through relatively passive measures rather than an all-out space weapons competition.

By racing to develop its own space weapons, the United States would cause two unfortunate sets of consequences. Militarily, it would legitimate a faster space arms race than is otherwise likely—something that can only hurt a country that effectively monopolizes military space activities today. Second, it would reinforce the current prevalent image of a unilateralist United States, too quick to reach for the gun and impervious to the stated will of other countries (as reflected in the huge majority votes at the United Nations in favor of negotiating bans on space weaponry). Among its other implications, this perception can make it harder for the United States to oppose treaties that it has good reasons to oppose—as was the case when the Bush administration withdrew from the ABM Treaty. It can also be harder for the United States to uphold international nonproliferation norms if its own actions weaken its credibility in demanding that others comply with arms control regimes.

Conversely, the support for wide-ranging bans on space weaponry in much of the arms control community is unjustified. Such accords would be generally unverifiable. They would also be incapable of changing the simple fact that many ballistic missile defense systems have inherent antisatellite potential that could allow them to be transformed into antisatellite weapons with relatively modest adjustments. This is a fact of physics, not of policy, and cannot be changed. A similar conclusion applies to many space technologies, including the increasingly prevalent microsatellites that are especially hard to monitor.

Nor can the United States permanently forswear the need for antisatellite capabilities. It does not need ASATs for now. That said, it will not realistically be able to continue its monopoly on the current array of space technologies, which allows it to use space assertively and confidently for military intelligence, communications, and tactical warfighting while its potential enemies cannot do so. And it needs to recognize that other countries are already interested in challenging America's military space monopoly, regardless of their political rhetoric on the subject. While most may be a ways from advanced dedicated ASAT systems—contradicting the sense of dire urgency of the 2001 Space Commission—they do already have nuclear-armed ballistic missiles and small laser devices with at least some inherent antisatellite potential.[2] So a moderate and nuanced policy, rather than an absolutist or ideological one, is the right path for the United States to follow.

Hardening and Defending (or Doing without) U.S. Satellites

How can military satellites be protected? And to the extent protections are insufficient, how can backups be developed for possible emergency use in war? The basic fact of the matter is that protection can be developed against a number of electronic threats, but explosives are difficult to counter. As such, satellite vulnerability is a physical fact of life. Moreover, the U.S. military's increasing dependence on commercial satellites for communications means that it is now vulnerable to relatively simple jamming as well.[3]

Several types of defensive responses can be imagined to counter a growing vulnerability of American satellites. At the simplest level, greater monitoring of space activities may be desirable so that the United States will know with more confidence if and when its satellites are being threatened. Greater hardening and other passive defenses—against nuclear effects, lasers and artificial heating, homing microsatellites, jamming, microwave weapons—would be next on the list. Then, some simple satellite defenses, such as greater fuel capacity for maneuvering and possible means of attacking homing enemy microsatellites, could be envisioned. Ultimately, if and when it was determined that all of the above could not reliably defend U.S. space assets, further measures may be needed—ranging from the capacity for rapid launching of replacement satellites to airborne substitutes for satellites. (ASATs are discussed later.)

The specific recommendations that emerge from this analysis are straightforward. First, military satellites should continue to be hardened against nuclear effects, and where practical, more should also employ radio transmission frequencies and signal strengths capable of penetrating a nuclear-disturbed atmosphere. These recommendations should be straightforward to implement; indeed, they already have been carried out for some systems, such as MILSTAR. These measures will ensure at least minimum levels of bandwidth even shortly after a nuclear attack, though not necessarily enough for most tactical warfighting purposes until the atmosphere begins to return to normal.

Second, low-Earth orbit satellites should have sensors capable of detecting laser illumination and possibly other attack mechanisms, as well as the means to protect themselves temporarily against such attacks through shutter controls that would shield their optics. (Someday, they may also need means of shielding themselves from prolonged exposure from high-energy lasers—Richard Garwin, for example, proposes a deployable spinning "parasol" with the shiny side facing Earth and the black side facing the satellite.) Such programs are reportedly gaining support at the Pentagon.

Third, despite such measures, it should be assumed that many types of military satellites may not be available in future war, and alternatives should therefore be maintained. This is particularly true for lower-altitude assets.

Fourth, plans should be made in the event that commercial communications satellites, which probably cannot be hardened in any practical way, prove unavailable for warfighting. That assumption should lead the U.S. military to devise means for making do with much-reduced bandwidth in combat; it should also buttress efforts to develop more dependable means of communication, such as laser satellite constellations.

Improved Space Monitoring

The United States needs to know if its satellites are under attack or likely soon to be under attack.[4] Otherwise, notice may only occur as multiple simultaneous satellite failures allow for no other real possibility. Sensors can trigger the deployment of shields or other protective measures against certain types of threats, such as jammers or lasers. They may allow for satellite maneuvers or other means of evading kinetic or explosive attack, as discussed below. For example, if an enemy ASAT were in reasonably close proximity, perhaps violating the boundaries of a treaty-defined keep-out zone, it might be defeated with high-energy short-range microwaves by a device that would not necessarily constitute a more general ASAT capability. But leaving aside the possible responses, which are not urgently needed at present, space awareness is important on several grounds and should be improved now.

Some U.S. satellites, including Defense Support Program early-warning assets and National Reconnaissance Office imaging satellites, already have some attack warning capability. But most U.S. satellites apparently do not.[5] The U.S. space surveillance network can track the movements of larger objects or boosters, and that may suffice against homing space mines for now. But at some future date, perhaps soon, satellites may need their own warning of approaching microsats. And low-altitude satellites should soon have sensors that would alert them to artificial illumination by laser.

Greater Resilience to Jamming

It is generally fairly easy to jam the communications links of satellites that have not been made resilient to such attacks. As one example, at the Air Force Research Laboratory, engineers "homebuilt" an effective jammer using about $7,500 worth of goods bought at electronics and hardware stores.[6] Similar devices have been built elsewhere, such as Schriever Air Force Base in Colorado. And Cuba recently interfered with the operations of a commercial communications satellite.[7]

A good deal of protection can be provided in this area, but it is unlikely to be affordable for commercial satellites, on which the U.S. military depends for many high-data-rate transmissions, such as those needed in tactical targeting (even if not for most high-level strategic command and control operations). Such measures can require much higher power and can reduce the flow of data substantially, fundamentally changing the economics of commercial communications. Among its other implications, that fact heightens the importance of moving along with the laser satellite communication system now under development by the Department of Defense, which will provide enormous bandwidth through the military's own system.

But the military also needs to prepare for the possibility that it may not have as much available communications bandwidth as it would like to have in future conflicts. The United States needs to ensure some level of robust, survivable satellite communications. New DSCS satellites with bandwidths in the vicinity of 60 Mbps are a step in the right direction (well above the MILSTAR capacities of 1 to 2 Mbps).[8]

Data transmission rates need to be minimized as much as possible. That can be accomplished through data compression techniques that can transmit high-fidelity data with one-tenth the bandwidth, or slightly degraded data at one-hundredth the bandwidth, of standard means.[9] It can also be done by maximizing the amount of analysis done by the platform obtaining the data.[10] Finally, the military needs to develop procedures for prioritizing its use of satellites, so that it can make do with less capacity if necessary. It may be necessary, for example, to reduce the use of video teleconferencing, and to downlink UAV live-feed imagery only to targeted receivers rather than throughout the battlefield.[11]

New GPS satellites with greater power will also be helpful to counter jamming and should not be again postponed (the GPS 3 constellation is to begin deployment in 2011, according to current plans).[12] If possible, indeed, deployment should be hastened. For now, inertial guidance or other terminal guidance may still be needed as a supplement to GPS for munitions used against a foe capable of jamming operations.

Improved Electronic Hardening

Hardening against nuclear effects can require a thorough shielding or redesign, to prevent an extremely intense and short electronic pulse from finding a "back door" into a satellite. Costs may grow by a few percent, up to perhaps 10 percent, as a result, but for military satellites in particular, this would hardly be onerous. If there has been any letup in such hardening since the cold war ended, it should be rectified; it is scarcely beyond the realm of the conceivable that an enemy would attack U.S. satellites with nuclear weapons.[13]

It is dubious that such hardening will ever be implemented for most commercial satellites, however, again underscoring the importance of not continuing to depend so greatly on such capabilities for wartime purposes. Even if the government were prepared to subsidize such hardening, commercial satellites would remain vulnerable to jamming and to direct attack, calling into question the value of the effort.

For military systems, however, hardening should be de rigeur. It is important for low-Earth orbit systems, given their proximity to Earth-based threats.[14] It is also desirable at higher altitudes. Satellites in MEO are often already hardened, since normal Van Allen radiation is greater at such altitudes. But standards may not be sufficiently demanding for all altitudes, from what can be deduced through unclassified sources. If true, that situation should be remedied.

There is yet another reason for radiation hardening, apart from nuclear threats. Within perhaps fifteen years, countries such as China could have the capacity to attack a variety of satellites using high-powered microwaves. The basic physics of radio-frequency weapons and high-powered microwave weapons is not particularly complicated. The engineering challenges associated with building devices that can emit very short pulses of radio energy lasting perhaps just billionths of a second but reaching billions of watts in power are considerable, but far from insurmountable.[15] As satellites are designed and produced in the coming years, such possible enemy capabilities should form part of the assumed future threat environment.[16] Satellites can be hardened against the electronic interferences created by microwave weapons, largely by shielding their soft electronic spots with a thin metal foil.

Increased Defenses against Explosives

Physically shielding satellites from the effects of explosives detonated as close as hundreds or dozens of meters, or even fewer, is difficult, given the ability of a hunter-killer satellite or space mine to approach arbitrarily close to a target satellite before being detonated. It is probably simply not worth the effort even to attempt such protection.

Could satellites maneuver, or be given self-defense weapons, to evade hunter-killer satellites? (Decoys would be difficult to employ effectively, given the size and electronic transmissions of large, modern satellites.) If a small satellite can devote most of its mass to fuel, it may be able to outmaneuver a large imaging satellite. As a general proposition, such maneuvering may be successful against simple ASATs with poor terminal guidance but is likely to fail against small sophisticated ASATs.[17] Perhaps a larger satellite could be equipped with a small short-range weapon to fire at such a device—something that would not create space debris. Such concepts should be researched, to see if defensive capabilities are feasible without at the same time developing weapons indistinguishable from offensive ASATs in their potential applications. Increased maneuvering capability may not be a permanent solution, but it could buy the United States time down the road and should be retained as an option, albeit a costly one, given the corresponding fuel requirements.

Backup Satellite Capabilities and Alternatives to Satellites

If the United States could take the expensive but prudent step of having some additional satellite capability in its inventory at all times, together with the ability to launch and make operational such satellites quickly, it would mitigate its vulnerability to antisatellite weapons. In particular, it would be better prepared against ASAT threats that were only capable of incapacitating a small number of its space assets.

Largely for this reason, Strategic Command would like to gain the capacity to replenish satellites in orbit within days. It hopes to have such an ability toward the end of the decade.[18] However, since that goal was articulated in 1998, the United States has not made rapid progress toward lowering launch or satellite costs.

Regardless of progress on the rapid relaunch front, the United States is probably entering an era when it should no longer count on its satellites remaining safe and secure. It is unlikely that any foe is close to having the ability to "clean up the heavens," systematically eliminating the dozens of GPS and communications satellites on hand for U.S. military use when needed. But satellites deployed now only in small numbers, such as imaging and signals intelligence satellites, may more plausibly be attacked. Over the long run, microsatellites or directed-energy weapons may even put the large constellations at risk. Although such a time is probably quite distant, the United States should avoid blind optimism about the continued availability of all its satellite capabilities.

As a practical matter, this conclusion has several implications. First, numerous airborne assets, particularly for imaging and signals intelligence, but also for targeting, guidance, and communications, should be in the force posture despite their nontrivial cost. In some cases, for assets such as P-3 aircraft and EC-135 electronic reconnaissance aircraft, refurbishment or modernization programs will be appropriate; in others, new and less expensive assets (largely UAVs) make more sense. Second, additional backup capabilities, such as fiber-optic land lines and undersea lines, should be retained in many regions of the world to permit high-volume intercontinental communications even if satellites are lost. Third, naval fleets, ground-force units, and aircraft should retain the ability to communicate internally through line-of-sight and airborne techniques, so that battle groups can always function as single entities, even if their access to satellites is disrupted.

The Offensive Option: Antisatellite Weapons

Despite the wide range of available policy options in the defensive realm, the United States may also need offensive military capabilities in space at some point. It does not need them now. But that could change.

Latent ASAT capabilities are already in the hands of many U.S. rivals and foes, primarily in the form of nuclear-tipped ballistic missiles. Many countries capable of space launch could also probably develop, in fairly short order, ASATs similar in principle to the Soviet co-orbital interceptor concept developed in the 1970s. To date they have not yet done so, as far as we know, though it is remotely possible that a country could test such a capability under the guise of putting a satellite into space (by trying to guide it to a moving aimpoint following the trajectory of a simulated satellite). Development of microsatellites may give countries other, somewhat stealthier, options as well, over time. This is not a trivial undertaking, but the technology is advancing and can be expected to keep doing so. Microsatellites might also be able to "sneak up" on other satellites gradually, using nitrogen jets rather than a rocket with its associated plume for the final approach.[19] Finally, ground-based directed-energy systems, such as high-energy lasers, may be of concern, too. All of these types of capabilities would be difficult to prohibit using arms control arrangements and standard verification tools. If other countries developed ASAT capabilities, a corresponding U.S. capability would probably be prudent as a deterrent, if for no other reason.

In addition, it is conceivable that the United States would wish to be the first to develop ASAT capabilities under certain circumstances in the future. Specifically, if an enemy could plausibly develop a war-winning capability, or even a notable military advantage, through use of its own satellites, the United States might decide that its security would be promoted by possessing ASATs. That might be true even if acquiring an ASAT spurred other countries to develop similar capabilities that put U.S. assets at risk.

As discussed in chapter 4, if in a future conflict near its shores China had imaging satellites capable of finding U.S. aircraft carriers and then passing targeting information to platforms carrying long-range antiship missiles, U.S. aircraft carriers might be put at acute risk. Oher countries might develop similar capabilities, too. ASATs might then be seen as the only way to make continued carrier operations in such waters feasible. Indeed, the United States might be willing to tolerate an ASAT arms competition in which its own satellites were put at greater risk in order to ensure incapacitation of the potential enemy's ability to strike large valuable American targets. This would be particularly true if the United States heeded the above advice about defensive measures and made sure its satellite capabilities were hardened, redundant, and backed up with nonorbiting assets that could take over the roles normally played by satellites if need be. In such circumstances, as the country projecting power, the United States might have a disproportionate dependence on large and vulnerable military assets; it would also probably have a greater ability to substitute other types of command, control, communications, and intelligence assets for satellites. So an ASAT competition might improve its prospects for decisive victory in such a war—and hence also improve its ability to deter the conflict in the first place—relative to an arrangement in which military space assets were left free to operate.

The Pros and Cons of Weaponization

The above discussion is not meant to sanction the development and use of antisatellite weapons now. The United States could never take the decision to engage in an ASAT competition lightly. Given the degree of international opposition to the weaponization of space, the potentially destabilizing effects of attacking satellites that provide reassurance and communications during crises, and the debris that could be created in orbital zones near Earth from kinetic energy and explosive weapons in particular (should other countries develop such weapons), ASATs would have major downsides. In addition, the United States benefits greatly from the status quo in space, in which it enjoys virtually exclusive capabilities to find and target enemy forces using satellite technology; it should try to preserve this state of affairs as long as possible. That the advantages of ASATs might outweigh these downsides at a future date is at least possible. But the time is not yet right for that approach.

A cautious military planner might naturally tend to disagree with the above assessment and advocate that the United States progress more rapidly toward putting various types of weapons in space. But cautious military planners should not make American security policy by themselves; their views should be balanced by those of cautious strategic planners. And the latter know that pursuit of unilateral military advantage sometimes leads to dynamics that can render one's own country, as well as the potential adversary, less secure. Examples abound in the realm of weapons of mass destruction—a much different arena from that of space warfare, but still illuminating as a precedent. The United States has in modern times elected not to pursue chemical or biological arms. It made that decision on the grounds that deploying such arms would likely reduce its own security—largely by legitimating weapons that the world community would be better off not to have widely used or even possessed, to the extent possible, especially given the potential for such weapons to fall into the hands of irresponsible and aggressive countries. It made similar decisions in the cold war in regard to missile defense, certain types of nuclear testing, nuclear weapons based in space, and indeed ASATs as well. In most cases, it did not doubt its ability to outcompete its potential adversary in narrow terms. But it recognized that the action-reaction, or arms race, dynamic that could well result would not advance its interests, and that in some cases it had asymmetric dependencies on assets such as satellites that argued for restraint in the development of weapons to threaten them.

Of course, many things have changed since the end of the cold war. But that fact argues for rethinking a number of American security policies from first principles, not for discarding them simply because they arose under different strategic circumstances.

A Hedging Strategy

So what are the proper components of U.S. strategy regarding the weaponization of space? What is a prudent hedging strategy? A central goal should be to make sure the United States is not surprised, and technologically outdistanced, by advances in ASAT capabilities that another country is able to achieve. A related goal should be to gradually explore technologies of potential use in ASATs, in case the United States someday finds it in its interest to develop these weapons—but not to actually develop any dedicated ASATs in the near-term future.

This philosophy argues for laboratory research on basic ASAT-related technologies. It also condones more advanced development and testing of systems with some inherent, if secondary, ASAT potential, such as the midcourse and airborne laser missile defense systems. But these should not now be given the final capabilities needed to work as ASATs (notably, means of finding and fixing on satellite targets) or tested in ASAT modes. Nor should an actual ASAT program be initiated anytime soon.

This approach could also involve some elements of formal arms control accords of indefinite duration. But any such limitations would have to be carefully defined and rather specific, most notably a ban on the creation of orbital debris. Informal restraint, perhaps through temporary and unilateral pledges, would be preferable in most cases. More specifically, elements of a "lead, but with restraint," or a hedging, strategy might include the following:

PURSUE LABORATORY RESEARCH AT A MODERATE LEVEL. It generally makes sense to conduct basic research and development. But this need not be overemphasized. Funding in the range of tens of millions of dollars a year for most basic types of ASAT technologies and concepts is adequate.

Because such indoor laboratory activities cannot be remotely monitored, and because they provide the United States long-term options it may someday need, they should not be disallowed by international accords, and the United States should pursue them itself. However, the scale of effort should be restrained, given that the need for ASAT-related technologies is not very urgent. Accelerating research now would waste money, risk sending the wrong message to other countries if and when the scale of a major program was revealed, and create bureaucratic and political pressures in the United States to ultimately field any system that was developed. None of those outcomes serve near- to medium-term U.S. security interests.

Overall, space-related R&D funding is robust now and need not be increased by more than already planned. Indeed, planned increases may be excessive in some cases, though it is difficult to be sure from unclassified sources. In 1999, space-related research accounted for about $432 million, or 39 percent of all Air Force science and technology funding. By 2005, these figures are expected to reach $847 million and 59 percent, respectively, and it is anticipated that they will keep going up thereafter.

The main drivers of these upward cost trends include laser communications, miniaturization concepts, imagery systems, and other satellite concepts ranging from ballistic missile defense to communications to navigation.[20] The Pentagon's February 2002 budget request included money for a number of broadly defined programs that may or may not have ASAT relevance and may or may not include more than basic scientific research: $40 million for directed energy technology, $14 million for space control technology, $65 million (in three different accounts) for high-energy laser research, and $122 million for ballistic missile defense technology in a part of the budget that had previously included funding for the space-based laser program. In the 2004 budget request, the Department of Defense requested about $250 million for very general space technology programs, about $85 million for high-energy laser research, and $15 million for space control technology. It also requested $82 million for "counterspace systems," a doubling from 2003, when they were a new budget item.[21]

More information and transparency from the Pentagon would be helpful for understanding the nature of these funds and the activities they support. In fact, there are reports that ASAT activity is further along than these numbers would suggest. For example, the Army has reportedly been working on laser dazzlers to blind surveillance satellites and jammers to disrupt communications and surveillance satellites.[22] The former, in particular, could be of concern if they are capable of causing lasting damage to other countries' satellites. And the Air Force reportedly has had similar efforts under way, which are likely to continue in some form—even if lethal ASATs may not be immediately pursued at the advanced development or testing stages.[23] The situation should be clarified by the Pentagon.

Any programs directly aimed at developing destructive ASAT weapons should be stopped. The airborne laser and midcourse ballistic missile defense systems already provide substantial hedges against the possible U.S. need for antisatellite systems. If the United States uncovers convincing, hard evidence that countries such as China are well along in ASAT development, it might be justified in proceeding with research and development too—though, depending on the state of China's efforts, not necessarily in testing.

CONTINUE ADVANCED DEVELOPMENT/DEPLOYMENT OF MISSILE DEFENSE CONCEPTS. Systems such as the Clinton administration's midcourse defense could easily have capabilities against low-altitude satellites, which move at roughly the altitudes and speeds characteristic of ballistic missile warheads. That system is nearly halfway through its scheduled flight test program and is slated for initial operational deployment on a provisional basis soon. Even today, a test missile and kill vehicle would probably have a respectable chance of destroying a satellite in LEO if first linked to the U.S. space surveillance system. In that sense, the United States already has a latent nonnuclear ASAT, even if its prototype missiles developed in the 1980s for launch from F-15s are no longer functional.

Other missile defense concepts may have similar capacity. Notable is the airborne laser, designed primarily to intercept relatively short-range missiles in their boost phase (see chapter 3). LEO satellites are probably no more difficult to reach with its beam than a burning rocket within the upper atmosphere. They would not be destroyed via the same mechanism as a liquid-fueled ballistic missile, the intended target of the ABL, but in many cases could be damaged or destroyed by its megawatt-class laser. The airborne laser is not quite as advanced as the Clinton midcourse system, but it could be capable of an intercept within two to three years, if the program stays on course.

The fact that these types of programs thus will provide real, if latent, ASAT capabilities rather soon is not a reason to cancel or curtail them. Missile defense is a sufficiently worthwhile enterprise to justify the effort. LEO satellite trajectories are so similar to those of ballistic missiles—in fact, easier to intercept, since they are more predictable—that a long-range midcourse missile defense system is in effect also a latent ASAT, at least within certain geographic constraints. Because other means of countering enemy satellites—jamming downlinks and uplinks, destroying ground stations, hiding U.S. military assets or making them hard to track—are not foolproof, some ASAT backup may prove prudent in the future. The possibility that the United States will someday need ASAT capability is great enough that missile defense systems with potential ASAT applicability are not undesirable. They are, in fact, a convenient hedge. At the same time, however, it is strongly preferable that they are not yet provided all the capabilities needed for ASAT purposes, or tested against satellites. An approach of hedging makes the most sense.

A Role for Arms Control?

Although the United States may need ASAT capability at some future date, certain restraints may be desirable. Some could be informal, some temporary. In general, they should be carefully tailored so as not to preclude development of various capabilities in the future. But they could still help reassure other countries about U.S. intentions at a time of still-unsettled great power relations and help protect space from the creation of excessive debris or other hazards to robust and safe use. Most of all, they could help protect military space assets on which the United States increasingly depends.

Some constraints might be formalized. For example, destructive testing of weapons, such as the Clinton midcourse missile defense system, against objects in satellite orbital zones would not only increase the risks of an ASAT competition. It would also create debris in LEO regions that would remain in orbit indefinitely (unless the testing occurred in what were effectively the higher parts of Earth's atmosphere, where air resistance would ultimately bring down debris and where few if any satellites fly in any case). A treaty banning any tests of ballistic missile systems or ASAT systems that would cause debris at altitudes above 300 miles makes sense.

Less formally, the United States should state that it will not test missile defense systems such as the ABL against objects in space for the foreseeable future, and that it will not create a real or virtual space-based test bed in the near future, either. Testing is not necessary to ensure the inherent ASAT capabilities of such systems. Tracking and pointing at satellite targets can be tested without firing weapons, so a system such as the ABL can confidently be assumed to have inherent ASAT capability without testing it in that mode.

Moreover, it is desirable to avoid the final steps of providing the ABL and other missile defense systems with an ASAT capability. It is better to show some level of restraint, even as the basic technological wherewithal for someday developing an ASAT is ensured. Moving quickly and explicitly to an ASAT capability would likely open a Pandora's box of international outcry and military and strategic responses at U.S. expense. Testing would only be needed at the final stage of weaponization; the United States is nowhere near that point. The hope is that it never will reach that point, if relations with the other great and spacefaring powers continue to improve. Policy should serve that latter goal rather than the narrow goal of rapidly maximizing ASAT capabilities on the assumption that the United States will fight countries such as China in the future. Such assumptions are unwarranted and do not serve U.S. interests; if given free rein, they can become self-fulfilling prophecies. Military planners must not be allowed to trump broader strategic planners in the American security debate.

As a matter of policy, the United States should also publicly state that it has no dedicated ASAT programs under way and no intention of initiating any for the foreseeable future. This policy may someday have to be reversed, but it would be beneficial in the short term.

Emphasize Nondestructive ASAT Concepts

To reduce the onus and negative symbolism of any ultimate development of ASAT technologies, the United States should focus preliminary laboratory research on technologies that would have the minimum destructive effect on any systems against which they were ultimately used. The goal of the United States should, where possible, be to avoid destroying satellites, even in a situation where some type of countersatellite capability is ultimately deemed necessary.[24] Not only kinetic or explosive destruction, but even permanent damage to satellite optics or electronics should be avoided if it proves possible to neutralize satellites in a more temporary and benign fashion during conflict. This approach is mirrored in U.S. military doctrine; the challenge is not in gaining theoretical consensus but in providing resources where appropriate.

Options should include jamming communications and destroying or otherwise neutralizing ground stations. The latter tactic was used in Operation Desert Storm and could be pursued in many cases, at least against countries possessing only fixed and known ground stations. Jamming will not necessarily work against a sophisticated adversary capable of frequency-hopping operations, but can generally be successful against less sophisticated adversaries. The United States is now looking into more deployable jammers that could deny adversaries the use of communications systems during conflict (by being located in the combat theater, near enemy lines of communication, such jammers could not easily be tuned out).[25] Over the longer term, high-powered microwaves could provide an option for either lethal effects (if used at maximum pulse power) or, better yet, temporary effects (if used at a lower, steadier power).[26]

In addition, the United States should explore other nonlethal ASAT concepts, such as devices launched into space that would unfurl large opaque shrouds just below enemy satellites, so that the latter could not track objects on Earth or communicate with Earth. Such options may not always be dependable or quickly usable in the event of a crisis or war, but they should be investigated and perhaps someday built if necessary. This could give the United States ASAT capabilities without clearly crossing the line toward space weaponization. But even these capabilities are not operationally required now; research and development efforts will suffice.

Conclusion

The United States depends enormously on its military space assets today. They do not function primarily as the great stabilizers and arms control facilitators of the cold war; in general, they have become the tools of the tactical warfighter. That reduces the strategic and political case for treating them as protected assets or viewing space as a sanctuary from military competition.

But any U.S. policy to pursue the actual weaponization of space in the near term would be a mistake. It would probably lead to an arms competition that would put American assets at risk sooner than they otherwise would be. Coming in the face of strong international opposition, it would further exacerbate the image of the United States as a go-it-alone power. That could, in turn, weaken Washington's ability to hold other countries to their arms control and nonproliferation commitments and to induce multilateral cooperation on other security issues.

That said, military space competition will occur regardless of American policy, and other countries will gradually learn to use space more like the United States does today. That calls for a two-tier approach from Washington. It must continue to anticipate, and protect against, attacks on its satellites to the extent possible. Commercial communications satellites and low-altitude military assets are probably the most apt to be vulnerable fairly soon, if not at present. Measures ranging from improved satellite hardening against lasers to more maneuvering capability against microsats to retention of ground-based alternatives to satellites may thus be required. In addition, without crossing the rubicon of weaponization or testing, the United States should keep its technology options open for the development of antisatellite weapons. Certain missile defense systems, together with laboratory research on basic technologies, already provide such inherent, latent capabilities. No dedicated ASAT programs are needed or desirable. And indeed the United States should say so, as a matter of national space doctrine for the foreseeable future.

The United States, leading the way on the increased militarization of space, may not be able to prevent the weaponization of space indefinitely. But slowing the process for as long as possible appears the best way to serve its core military and strategic interests.

NOTES

Notes to Chapter One

1. Paul B. Stares, *The Militarization of Space: U.S. Policy, 1945–1984* (Cornell University Press, 1985), pp. 62–71; and John Lewis Gaddis, *The Long Peace: Inquiries into the History of the Cold War* (Oxford University Press, 1987), pp. 195–214.

2. Peter L. Hays, *United States Military Space: Into the Twenty-First Century* (Montgomery, Ala.: Air University Press, 2002), pp. 15–20. The full names of these agreements are the Anti-Ballistic Missile (ABM) Treaty, the Strategic Arms Limitation Talks (SALT), the Conventional Armed Forces in Europe (CFE) Treaty, and the Strategic Arms Reduction Treaty (START).

3. Department of Defense, *Conduct of the Persian Gulf War: Final Report to Congress* (1992), p. K-41.

4. Jason Bates and Jeremy Singer, "GPS Devices Proving Key to Avoiding Fratricide," *Space News*, September 15, 2003, pp. 25–27.

5. Barry D. Watts, *The Military Uses of Space: A Diagnostic Assessment* (Washington: Center for Strategic and Budgetary Assessments, 2001), pp. 44–45.

6. See Thomas A. Keaney and Eliot A. Cohen, *Gulf War Air Power Survey Summary Report* (Government Printing Office,

1993), p. 193; Department of Defense, *Kosovo/Operation Allied Force After-Action Report* (2000), p. 46.

7. William B. Scott, "Milspace Comes of Age in Fighting Terror," *Aviation Week and Space Technology*, April 8, 2002, pp. 77–78. About half of this bandwidth, or 300 million bits per second, was in the form of leased commercial capacity.

8. See Patrick Rayerman, "Exploiting Commercial SATCOM: A Better Way," *Parameters* (Winter 2003–04), p. 55.

9. Michael Sirak, "Flexibility Key to Weapon Mix," *Jane's Defence Weekly*, June 18, 2003, p. 45; Andy Pasztor, "France's Eutelsat Hits Jackpot with U.S. Satellite Contracts," *Wall Street Journal*, March 28, 2003; Vernon Loeb, "Intense, Coordinated Air War Backs Baghdad Campaign," *Washington Post*, April 6, 2003, p. 24; and J. R. Wilson, "Plying the Space-Based Advantage," *Armed Forces Journal* (June 2003), pp. 40–42.

10. William B. Scott, "Milspace Will Be a Major Player in 'Gulf War 2,'" *Aviation Week and Space Technology*, January 13, 2003, p. 398.

11. Theresa Hitchens, "Monsters and Shadows: Left Unchecked, American Fears Regarding Threats to Space Assets Will Drive Weaponization," *Disarmament Forum*, vol. 1 (2003), p. 17.

12. See Congressional Research Service, "U.S. Space Programs," CRS Issue Brief IB 92011 (September 16, 2003), Summary.

13. Personal communication from General Mike Hamel, 14th Air Force, Vandenberg Air Force Base, California, November 8, 2002.

14. Benjamin S. Lambeth, *Mastering the Ultimate High Ground: Next Steps in the Military Uses of Space* (Santa Monica, Calif.: RAND, 2003), p. 90.

15. Ian Steer and Melanie Bright, "Blind, Deaf, and Dumb," *Jane's Defence Weekly*, October 23, 2002, pp. 20–23.

16. Paul B. Stares, *Space and National Security* (Brookings, 1987), pp. 85–113.

17. Ashton B. Carter, "Satellites and Anti-Satellites: The Limits of the Possible," *International Security*, vol. 10, no. 4 (Spring 1986), p. 75.

18. Notable works on space-based ballistic missile defense, as well as antisatellite technology, include Carter, "Satellites and Anti-Satellites;" Fred S. Hoffman, "Ballistic Missile Defenses and U.S. National Security," in Harold Brown and others, eds., *The Strategic*

Defense Initiative (Claremont, Calif.: Keck Center for International Strategic Studies, 1985); Ashton B. Carter, *Directed Energy Missile Defense in Space*, Background Paper (Washington: Office of Technology Assessment, 1984); Stares, *Space and National Security*; the Pentagon's yearly *Soviet Military Power*; John Tirman, ed., *The Fallacy of Star Wars* (Vintage Books, 1984); and James T. Fletcher, *The Strategic Defense Initiative Defensive Technologies Study* (Department of Defense, 1984).

19. See Stares, *Space and National Security,* p. 147.

20. Peter L. Hays, "Military Space Cooperation: Opportunities and Challenges," in James Clay Moltz, ed., *Future Security in Space: Commercial, Military, and Arms Control Trade-Offs*, Occasional Paper 10 (Monterey, Calif.: Monterey Institute of International Studies, 2002), p. 37.

21. This view is hardly confined to conservatives; see, for example, Carter, "Satellites and Anti-Satellites," p. 47.

22. Jonathan Dean, "Defenses in Space: Treaty Issues," in Moltz, *Future Security in Space,* p. 4.

23. See William B. Scott, "ASAT Test Stalled by Funding Dispute," *Aviation Week and Space Technology*, July 1, 1996, p. 59; Greg Caires, "Limited ASAT System Could Be Fielded by 1999, Army Says," *Defense Daily*, August 5, 1996; Hunter Keeter, "Kinetic Energy Anti-Satellite System Hangs in the Balance," *Defense Daily*, June 29, 1999; Colonel John E. Hyten, "A Sea of Peace or a Theater of War?" *Air and Space Power Journal* (Fall 2002), p. 81; and George C. Wilson, "Mr. Smith's Crusade," *National Journal*, vol. 33, no. 32 (August 11, 2001).

24. Steven Lambakis, *On the Edge of Earth* (University Press of Kentucky, 2001), p. 102.

25. Benjamin S. Lambeth, *Mastering the Ultimate High Ground* (Santa Monica, Calif.: RAND, 2003), p. 88.

26. Michael Krepon with Christopher Clary, *Space Assurance or Space Dominance? The Case Against Weaponizing Space* (Washington, D.C.: Henry L. Stimson Center, 2003), p. 21.

27. Donald Rumsfeld, "Report of the Commission to Assess United States National Security Space Management and Organization" (Washington, January 11, 2001).

28. Secretary of Defense Donald H. Rumsfeld, *Quadrennial Defense Review Report 2001* (Department of Defense, 2001), p. 45.

29. Colonel Frank G. Klotz, *Space, Commerce, and National Security* (New York: Council on Foreign Relations, 1998); and Hays, *United States Military Space*, pp. 15–20.

30. U.S. Space Command (before merging with Strategic Command), *Long Range Plan* (1998), chap. 5, available at www.spacecom.mil/LRP.

31. See the White House, "Fact Sheet, National Space Policy," September 19, 1996, available at www.aiaa.org/policy.

32. John A. Tirpak, "Challenges Ahead for Military Space," *Air Force Magazine* (January 2003), p. 25.

33. U.S. Space Command, *Long-Range Plan,* pp. 20–48.

34. See Lt. Col. Larry J. Schaefer, *Sustained Space Superiority: A National Strategy for the United States*, Occasional Paper 30 (Montgomery, Ala.: Air University Press, 2002); Hyten, "A Sea of Peace or a Theater of War?" pp. 80–86; Major John Grenier, "A New Construct for Air Force Counterspace Doctrine," *Air and Space Power Journal* (Fall 2002), pp. 20–21; and M. V. Smith, "Ten Propositions Regarding Spacepower," Thesis, School of Advanced Airpower Studies, Maxwell Air Force Base, Ala., June 2001.

35. Jeremy Singer, "Pentagon Renews Interest in Small Launchers," *Space News*, March 3, 2003, p. 3.

36. This change was effective October 1, 2002. See William B. Scott, "'New' Strategic Command Could Assume Broader Roles," *Aviation Week and Space Technology*, October 14, 2002, p. 63. The Air Force still has a service-specific Space Command with 25,000 active-duty military personnel and civilians as well as 14,000 contractors; see Air Force Space Command, "Fact Sheet," January 2002, www.af.mil/news/factsheets/Air_Force_Space_Command.html.

37. See James Dao, "Rumsfeld Plan Skirts a Call for Stationing Arms in Space," *New York Times*, May 9, 2001, p. A13.

38. See, for example, Hitchens, "Monsters and Shadows," p. 24.

39. Hays, *United States Military Space*, pp. 18–25.

40. Watts, *The Military Uses of Space,* pp. 8, 15, 21.

41. See NGO Committee on Disarmament, "Panel Discussion Held in the United Nations," October 19, 2000, transcript, available at www. igc.org/disarm/T191000outerspace.htm; Hu Xiaodi, Ambassador for Disarmament Affairs of China, Statement at the Plenary of the Conference on Disarmament, June 7, 2001, available at www3.itu. int/missions/China/disarmament/2001files/disarmdoc010607.htm; and

"China, Russia Want Space Weapons Banned," *Philadelphia Inquirer*, August 23, 2002.

42. See "Canadian Working Paper Concerning CD Action on Outer Space," January 21, 1998, available at www.unorg.ch/disarm/curdoc/1487.htm; and James Clay Moltz, "Breaking the Deadlock on Space Arms Control," *Arms Control Today* (April 2002), available at www.armscontrol.org/act/2002_04/moltzapril02.asp?print.

43. Dean, "Defenses in Space," p. 5.

44. Vitaly A. Lukiantsev, "Enhancing Global Security through Improved Space Management: A Russian Perspective," in Moltz, *Future Security in Space*, p. 47; Cheng Jingye, "Treaties as an Approach to Reducing Space Vulnerabilities," in Moltz, *Future Security in Space*, p. 49; and Dean, "Defenses in Space," p. 5.

45. Hays, *United States Military Space,* pp. 11–13; Alvin and Heidi Toffler, *War and Anti-War: Survival at the Dawn of the 21st Century* (Boston: Little, Brown, 1993); Stuart E. Johnson and Martin C. Libicki, eds., *Dominant Battlespace Knowledge* (Washington: National Defense University Press, 1996); Keaney and Cohen, *Gulf War Air Power Survey Summary Report*; William Owens, *Lifting the Fog of War* (Farrar, Straus, and Giroux, 2000); Daniel Goure and Christopher M. Szara, eds., *Air and Space Power in the New Millennium* (Washington: Center for Strategic and International Studies, 1997); Defense Science Board 1996 Summer Study Task Force, *Tactics and Technology for 21st Century Military Superiority* (Department of Defense, 1996); Harlan Ullman and others, *Shock and Awe: Achieving Rapid Dominance* (Washington: National Defense University Press, 1996); George Friedman and Meredith Friedman, *The Future of War: Power, Technology, and American World Dominance in the 21st Century* (Crown, 1996); John Arquilla and David Ronfeldt, eds., *In Athena's Camp: Preparing for Conflict in the Information Age* (Santa Monica, Calif.: RAND, 1997); National Defense Panel, *Transforming Defense: National Security in the 21st Century* (Arlington, Va.: December 1997); and Joint Chiefs of Staff, *Joint Vision 2010* and *Joint Vision 2020* (Department of Defense, 2000 and 2002).

46. Watts, *The Military Uses of Space,* pp. 29–30.

47. Bill Gertz, "Chinese Missile Has Twice the Range U.S. Anticipated," *Washington Times*, November 20, 2002, p. 3.

48. "Report of the Commission to Assess United States National Security Space Management and Organization" (Washington, January 11, 2001), pp. 22–23.

49. Although there has been a recent resurgence of publications about the future military uses of space, there has not yet been a study that takes my approach, focusing squarely on future space policy in a pragmatic way that attempts to balance technical and strategic considerations. Two very good primers tend to shy away from policy analysis: Watts, *The Military Uses of Space*; and Bob Preston and others, *Space Weapons, Earth Wars* (Santa Monica, Calif.: RAND, 2002). Lambakis, *On the Edge of Earth,* provides good background material as well. He finishes with strong advocacy for U.S. weaponization of space, a position derived principally from a realist and determinist sense about the nature of human technology and human conflict. Since space is a new arena in which competition is possible, he argues, such competition is inevitable; the choice for the United States is whether to lead in that competition or to lose it. The argument is initially plausible, but also rather categorical, and in the end not persuasive.

50. See also James M. Lindsay and Michael E. O'Hanlon, *Defending America: The Case for Limited National Missile Defense* (Brookings, 2001).

Notes to Chapter Two

1. Ashton B. Carter, "Satellites and Anti-Satellites: The Limits of the Possible," *International Security*, vol. 10, no. 4 (Spring 1986), pp. 50–52.

2. M. V. Smith, "Ten Propositions Regarding Spacepower," Thesis, School of Advanced Airpower Studies, Maxwell Air Force Base, Ala., June 2001, p. 44.

3. Escape velocity for an object at or near the surface of Earth is about eleven kilometers per second. See among physics textbooks on mechanics, for example, Robert Resnick and David Halliday, *Physics, Volume I* , 3d ed. (John Wiley, 1977), pp. 174–80; Herbert Goldstein, *Classical Mechanics* (Reading, Mass.: Addison-Wesley, 1980), pp. 29–31; and Curtis D. Cochran, Dennis M. Gorman, and Joseph D. Dumoulin, eds., *Space Handbook*, 12th rev. (Montgomery, Ala.: Air University Press, 1985), pp. 2-1–3-50.

4. Cochran, Gorman, and Dumoulin, *Space Handbook*, pp. 3–23. To see why, think of the GEO rocket as operating in two separate phases. First, it must reach the speed of the LEO rocket. At that point, it must have enough fuel left to propel the payload up to GEO. That

means it must accelerate not only the payload but also a large amount of fuel up to LEO speeds. Since that fuel has a large mass, the rocket must do far more work for the GEO payload. In contrast, for many vehicles operating on the surface of Earth extra mass imposes a much less severe penalty, since they are not directly fighting gravity in the way that a vertically oriented rocket must.

5. Barry D. Watts, *The Military Uses of Space: A Diagnostic Assessment* (Washington: Center for Strategic and Budgetary Assessments, 2001), p. 123.

6. See, for example, Thomas B. Cochran, William M. Arkin, and Milton M. Hoenig, *Nuclear Weapons Databook, Volume I: U.S. Nuclear Forces and Capabilities* (Cambridge, Mass.: Ballinger, 1984), pp. 1116–145; and Aviation Week and Space Technology, *2002 Aerospace Source Book* (January 14, 2002), pp. 144–53. Nor do trends in rocket propellants or rocket structural materials promise radical changes in this situation anytime soon; see Watts, *The Military Uses of Space*, p. 58.

7. Watts, *The Military Uses of Space*, p. 123.

8. See, for example, *Air Force Magazine* (August 2002), pp. 40–42.

9. Cochran, Gorman, and Dumoulin, *Space Handbook*, pp. 1-9–1-10.

10. U.S. Space Command, "Fact Sheet: Space Surveillance," February 2001, available at www.spacecom.mil/space.htm.

11. Marco Antonio Caceres, "Cutbacks Reflect Sluggishness in Commercial Satellite Market," *Aviation Week and Space Technology*, January 13, 2003, p. 151.

12. Marco Antonio Caceres, "Launch Services Market Going Nowhere Fast," *Aviation Week and Space Technology*, January 14, 2002, pp. 139–40; Marco Antonio Caceres, "Hopes Fading for Most RLVs," *Aviation Week and Space Technology*, January 14, 2002, p. 142; Marco Antonio Caceres, "Satellite Industry Stalls in Standby Mode," *Aviation Week and Space Technology*, January 14, 2002, p. 155; and Watts, *The Military Uses of Space*, pp. 15, 61.

13. Watts, *The Military Uses of Space*, p. 50.

14. Personal communication from Major David Outlaw, 14th Air Force, Vandenberg Air Force Base, California, November 8, 2002. The medium-secure data rate for MILSTAR (Military Strategic and Tactical Relay) satellites is about 1.5 million bits per second (1.5 megabits per second, or Mbps), and for each defense satellite communications system (DSCS) satellite about 60 Mbps; by contrast, as noted in chapter 1,

total bandwidth from all sources in Operation Enduring Freedom approached 1 billion bits per second (1 gigabit per second, or Gbps).

15. Joseph C. Anselmo, "Shutter Controls: How Far Will Uncle Sam Go?" *Aviation Week and Space Technology*, January 31, 2000, pp. 55–56.

16. Peter L. Hays, "Military Space Cooperation: Opportunities and Challenges," in James Clay Moltz, ed., *Future Security in Space: Commercial, Military, and Arms Control Trade-Offs*, Occasional Paper 10 (Monterey, Calif.: Monterey Institute of International Studies, 2002), p. 38.

17. Joel R. Primack, "Debris and Future Space Activities," in Moltz, *Future Security in Space,* p. 18.

18. Watts, *The Military Uses of Space*, p. 79.

19. Peter L. Hays, *United States Military Space: Into the Twenty-First Century* (Montgomery, Ala.: Air University Press, 2002), p. 133.

20. Primack, "Debris and Future Space Activities," p. 20.

21. Watts, *The Military Uses of Space*, pp. 42–43.

22. Jeffrey T. Richelson, *The U.S. Intelligence Community*, 2d ed. (Cambridge, Mass.: Ballinger, 1989), p. 201.

23. Craig Covault, "Secret NRO Recons Eye Iraqi Threats," *Aviation Week and Space Technology*, September 16, 2002, p. 23; Jeffrey T. Richelson, *America's Secret Eyes in Space: The U.S. Keyhole Spy Satellite Program* (Harper and Row, 1990), pp. 130–32, 186–87, 206–08, 227, 236–38; and Office of Technology Assessment, *Launch Options for the Future* (Washington, 1988), pp. 65–66.

24. Covault, "Secret NRO Recons Eye Iraqi Threats."

25. *Air Force Magazine* (August 2002), pp. 40–42; Carter, "Satellites and Anti-Satellites," p. 56.

26. George Friedman and Meredith Friedman, *The Future of War* (Crown, 1996), pp. 321–24.

27. James Bamford, *Body of Secrets: Anatomy of the Ultra-Secret National Security Agency* (Doubleday, 2001), p. 369.

28. Jason Bates, "U.S. Government to Broaden Use of Commercial Imagery," *Space News*, May 19, 2003, p. 6.

29. Jason Bates, "Military Helps Stabilize Satellite Imagery Market," *Space News*, August 25, 2003, p. 8.

30. Dan Cragg, *Guide to Military Installations*, 5th ed. (Mechanicsburg, Pa.: Stackpole Books, 1997), pp. 30, 57; and Richelson, *The U.S. Intelligence Community*, pp. 175–76.

31. Paul B. Stares, *Space and National Security* (Brookings, 1987), p. 188.

32. See the Air Force Space Command website, www.spacecom. af.mil/hqafspc/Library/Units/worldsites.htm; and the Federation of American Scientists website, www.fas.org.

33. Bruce G. Blair, *Strategic Command and Control: Redefining the Nuclear Threat* (Brookings, 1985), p. 253.

34. Watts, *The Military Uses of Space*, p. 78.

35. Benjamin S. Lambeth, *Mastering the Ultimate High Ground: Next Steps in the Military Uses of Space* (Santa Monica, Calif.: RAND, 2003), p. 136; and Robert S. Dudney, "Space Power in the Gulf," *Air Force Magazine* (June 2003), p. 2.

36. Craig Covault, "Military Satcom, Relay Programs Boost Industry, Enhance Warfare," *Aviation Week and Space Technology*, January 6, 2003, p. 43.

37. J. C. Toomay, *Radar Principles for the Non-Specialist*, 2d ed. (Mendham, N.J.: SciTech Publishing, 1998), p. 175.

38. Warren Ferster, "MDA to Buy at Least One More Missile Tracking Satellite," *Space News*, November 17, 2003, p. 3.

39. Lambeth, *Mastering the Ultimate High Ground*, p. 145.

40. At an altitude of 500 kilometers, it would require sixty satellites in polar orbit to maintain nearly continuous coverage. At 1,000 kilometers, that number would drop to about thirty; at 1,600 kilometers, it would be twenty. See Stares, *Space and National Security*, p. 40, based on Patrick J. Friel, "New Directions for the U.S. Military and Civilian Space Programs," in Uri Ra'anan and Robert L. Pfaltzgraff Jr., eds., *International Security Dimensions of Space* (Hamden, Conn.: Archon Books, 1984), p. 124, and D. C. Beste, "Design of Satellite Constellations for Optimal Continuous Coverage," *IEEE Transactions on Aerospace and Electronic Systems*, no. 3 (May 1978), pp. 466–73.

41. Watts, *The Military Uses of Space*, p. 80.

42. Michael Krepon with Christopher Clary, *Space Assurance or Space Dominance?: The Case against Weaponizing Space* (Washington: Henry L. Stimson Center, 2003), p. 23.

43. *Air Force Magazine* (August 2002), pp. 26–46; Aviation Week and Space Technology, *2002 Aerospace Source Book*, pp. 144–52.

44. *Air Force Magazine* (August 2002), pp. 26–46.

45. Watts, *The Military Uses of Space*, pp. 36–40.

46. Steven Lambakis, *On the Edge of Earth: The Future of American Space Power* (University Press of Kentucky, 2001), pp. 159–60.

47. Aviation Week and Space Technology, *2002 Aerospace Source Book*, pp. 161–73; and Associated Press, "China Launches Key Satellite," *Washington Post*, May 26, 2003, p. A23.

48. Aviation Week and Space Technology, *2002 Aerospace Source Book*, pp. 146–47.

49. Craig Covault, "Chinese Rocket R&D Advances," *Aviation Week and Space Technology*, November 12, 2001, pp. 54–55.

50. Department of Defense, *Annual Report on the Military Power of the People's Republic of China* (July 2002), pp. 28–29, 45; Lambakis, *On the Edge of Earth*, pp. 147–50.

51. See Eugene Gholz, "Military Transformation, Political Economy Pressures and the Future of Trans-Atlantic National Security Space Cooperation," *Astropolitics*, vol. 1, no. 2 (Autumn 2003), p. 32.

52. Aviation Week and Space Technology, *2002 Aerospace Source Book*, pp. 161–73.

53. Peter B. de Selding, "Europe Expected to Sign Galileo Contract by July," *Space News*, June 2, 2003, p. 7; and Laurence Nardon, "The World's Space Systems," *Disarmament Forum*, vol. 1 (2003), p. 37.

54. David Braunschvig, Richard L. Garwin, and Jeremy C. Marwell, "Space Diplomacy," *Foreign Affairs*, vol. 82, no. 4 (July–August 2003), pp. 156–64.

55. Aviation Week and Space Technology, *2002 Aerospace Source Book*, pp. 139–52; and "Sea Launch Information," www.sea-launch.com/special/sea-launch/information.htm.

56. Paul Kallender, "Spy Satellite Launch Marks New Era for Japan in Space," *Space News*, March 31, 2003, p. 8.

57. Barbara Opall-Rome, "Israel Moving Ahead with Military Satcom Plans," *Space News*, June 23, 2003, p. 8.

58. Barbara Opall-Rome and Vivek Raghuvanshi, "India Seeks Israeli Ofeq 5 Spy Satellite Imagery," *Space News*, September 22, 2003, p. 8.

59. Lambakis, *On the Edge of Earth*, pp. 146–62.

60. Watts, *The Military Uses of Space*, pp. 65–69; and Orbital Sciences' website, www.orbimage.com.

61. Joseph C. Anselmo, "Commercial Space's Sharp New Image," *Aviation Week and Space Technology*, January 31, 2000, p. 54.

62. Lambakis, *On the Edge of Earth*, p. 157.

Notes to Chapter Three

1. For an endorsement of such ideas, see Commission on the Future of the United States Aerospace Industry, *Final Report* (November 2002), p. 3-5, www.aerospacecommission.gov/Aero CommissionFinalReport.pdf.

2. Anne Marie Squeo, "Officials Say Space Programs Face Delays, Are 'in Trouble,'" *Wall Street Journal*, December 2, 2002, p. 1.

3. Bruce G. Blair, *Strategic Command and Control: Redefining the Nuclear Threat* (Brookings, 1985), pp. 201–07.

4. Ian Steer and Melanie Bright, "Blind, Deaf, and Dumb," *Jane's Defence Weekly*, October 23, 2002, pp. 21–23.

5. Blair, *Strategic Command and Control*, p. 206.

6. Donald Rumsfeld, "Report of the Commission to Assess United States National Security Space Management and Organization" (Washington, January 11, 2001), pp. 21–22.

7. Dennis Papadopoulos, "Satellite Threat Due to High Altitude Nuclear Detonations," briefing slides presented December 17, 2002, based on research done at the University of Maryland for Advanced Power Technologies, Inc. Cited here by the permission of the author.

8. Ashton B. Carter, "Satellites and Anti-Satellites: The Limits of the Possible," *International Security*, vol. 10, no. 4 (Spring 1986), pp. 89–92.

9. Barry D. Watts, *The Military Uses of Space: A Diagnostic Assessment* (Washington: Center for Strategic and Budgetary Assessments, 2001), p. 99.

10. Peter L. Hays, *United States Military Space: Into the Twenty-First Century* (Montgomery, Ala.: Air University Press, 2002), pp. 121–24.

11. Ira W. Merritt, U.S. Army Space and Missile Defense Command, "Radio Frequency Weapons and Proliferation: Potential Impact on the Economy," statement before the Joint Economic Committee, 105 Cong. 2 sess., February 25, 1998; Barbara Starr, "Russian Bomb-Disarming Device Triggers Concerns," *Jane's Defence Weekly*, March 18, 1998, p. 4; Carlo Kopp, "The E-Bomb—A Weapon of Electrical Mass Destruction" (Monash University, Australia, 1998); "Russian Electronic Bomb Tested in Sweden," *Agence France-Presse*, January 21, 1998; David A. Fulghum, "Microwave Weapons Await a Future War," *Aviation Week and Space Technology*, June 7, 1999, pp. 30–31;

and Lexington Institute, *Directed-Energy Weapons: Technologies, Applications, and Implications* (Washington, 2003), pp. 15–18.

12. Mark E. Rogers, *Lasers in Space: Technological Options for Enhancing U.S. Military Capabilities*, Occasional Paper 2 (Montgomery, Ala.: Center for Strategy and Technology, Air War College, November 1997), p. 18 of 86 (www.milnet.com/milnet/pentagon/laser/occppr02.htm).

13. Rogers, *Lasers in Space*, p. 16.

14. See Air Force Scientific Advisory Board, *New World Vistas: Directed Energy Volume* (Washington: U.S. Air Force, December 1995), as quoted in Rogers, *Lasers in Space*, p. 56.

15. Defense Science Board, *High Energy Laser Weapon Systems Applications* (Office of the Under Secretary of Defense for Acquisition, Technology, and Logistics, June 2001), p. 49.

16. High Energy Laser Executive Review Panel, "Department of Defense Laser Master Plan" (Office of the Deputy Undersecretary of Defense for Science and Technology, March 24, 2000), pp. 4–6.

17. Breck Hitz, J. J. Ewing, and Jeff Hecht, *Introduction to Laser Technology*, 3d ed. (New York: IEEE Press, 2001), p. 191; and David A. Fulghum, "Lasers, HPM Weapons Near Operational Status," *Aviation Week and Space Technology*, July 22, 2002 (www.aviationnow.com/content/publication/awst/20020722/aw173.htm).

18. Elihu Zimet, "High-Energy Lasers: Technical, Operational, and Policy Issues," *Defense Horizons* 18 (Washington: National Defense University, Center for Technology and National Security Policy, October 2002), pp. 6–7 of 16 (www.ndu.edu/inss/DefHor/DH18/DH_18. htm).

19. Defense Science Board, *High Energy Laser Weapon Systems Applications,* p. 54.

20. High Energy Laser Executive Review Panel, "Department of Defense Laser Master Plan," pp. 7, 10.

21. Zimet, "High-Energy Lasers," p. 6; Rogers, *Lasers in Space*, p. 56; and Defense Science Board, *High Energy Laser Weapon Systems Applications*, p. 49.

22. Geoffrey E. Forden, "The Airborne Laser," *IEEE Spectrum* (September 1997), pp. 47–49.

23. Robert E. Levin, Director, Acquisition and Sourcing Management, Testimony before the Subcommittee on National Security, Veterans' Affairs, and International Relations, Committee on Government Reform, House of Representatives, in General Accounting Office,

"Missile Defense: Knowledge-Based Process Would Benefit Airborne Laser Decision-Making," GAO-02-949T (July 16, 2002).

24. Forden, "The Airborne Laser," pp. 40–49; General Accounting Office, *Defense Acquisitions: DoD Efforts to Develop Laser Weapons for Theater Defense*, GAO/NSIAD-99-50 (March 1999); Missile Defense Agency, "Airborne Laser Completes First Flight," News Release, July 18, 2002; and Federation of American Scientists, "Special Weapons Monitor: Airborne Laser," updated September 18, 2002, available at www.fas.org/spp/starwars/program/abl.htm.

25. Zimet, "High-Energy Lasers," p. 6.

26. James Clay Moltz, "Reining in the Space Cowboys," *Bulletin of the Atomic Scientists* (January–February 2003), p. 63.

27. Select Committee on U.S. National Security and Military/Commercial Concerns with the People's Republic of China, House of Representatives, *U.S. National Security and Military/Commercial Concerns with the People's Republic of China* (1999), p. 209.

28. Department of Defense, *Annual Report on the Military Power of the People's Republic of China* (2002), pp. 5, 32, 40.

29. Zimet, "High-Energy Lasers," p. 6; Sandra I. Erwin, "Killing Missiles from Space: Can the U.S. Air Force Do It with Lasers?" *National Defense Magazine* (June 2001), p. 5 (www. nationaldefense magazine.org/article.cfm?Id=513).

30. Erwin, "Killing Missiles from Space," p. 5.

31. Defense Science Board, *High Energy Laser Weapon Systems Applications*, p. 3; and Missile Defense Agency, "Space Based Laser Fact Sheet," January 2002, available at www.acq.osd.mil/bmdo/bmdolink.

32. Erwin, "Killing Missiles from Space," p. 3.

33. Defense Science Board, *High Energy Laser Weapon Systems Applications*, pp. 16–31.

34. General Accounting Office, *Defense Acquisitions*.

35. Zimet, "High-Energy Lasers," p. 10.

36. Defense Science Board, *High Energy Laser Weapon Systems Applications*, pp. 19–31.

37. Celeste Johnson and Raymond Hall, *Estimated Costs and Technical Characteristics of Selected National Missile Defense Systems* (Congressional Budget Office, 2002), pp. 20–27.

38. As an indication of how seriously it takes this technology, the Pentagon requested $452 million in its 2004 budget proposal for

development of a laser satellite communications capability. It also requested $778 million for the advanced EHF satellite system (and $299 million for a new space-based radar). See Department of Defense, "FY 2004 Defense Budget Briefing Slides," February 3, 2003, p. 18, available at www.defenselink.mil.

39. It may be possible to transmit an image using one-tenth the amount of bandwidth currently employed, and if some fidelity (1 to 2 percent of pixels) can be sacrificed, it may be possible to transmit images with less than one-hundredth the current number of bytes. See Michael Sirak, "U.S. Air Force Targets UAV Bandwidth Problem," *Jane's Defence Weekly*, July 31, 2002, p. 28.

40. Peter L. Hays, "Military Space Cooperation: Opportunities and Challenges," in James Clay Moltz, ed., *Future Security in Space: Commercial, Military, and Arms Control Trade-Offs*, Occasional Paper 10 (Monterey, Calif.: Monterey Institute of International Studies, 2002), p. 41.

41. J. C. Toomay, *Radar Principles for the Non-Specialist*, 2d ed. (Mendham, N.J.: Scitech Publishing, 1998), p. 154.

42. Frank Tiboni, "U.S. Military Has Enough Bandwidth for War with Iraq, Pentagon CIO Says," *DefenseNews.com*, January 22, 2003.

43. Katie Walter, "Laser Zaps Communication Bottleneck," *Lawrence Livermore National Laboratory Science and Technology Review* (December 2002), pp. 18–21.

44. Craig Covault, "Military Satcom, Relay Programs Boost Industry, Enhance Warfare," *Aviation Week and Space Technology*, January 6, 2003, p. 43; and Rogers, *Lasers in Space*, pp. 45–46.

45. Amy Butler, "Wolfowitz Boosts MILSATCOM by Billions, Outlines Additional Buys," *Inside the Air Force*, December 20, 2002, p. 1.

46. Brian Berger, "SLI Overhaul Marks Major Shift in NASA Spending," *Space News*, November 18, 2002, p. 3.

47. Commission on the Future of the United States Aerospace Industry, *Final Report* (November 2002), p. 3-2.

48. Michael O'Hanlon, *Technological Change and the Future of Warfare* (Brookings, 2000), pp. 76–78.

49. Craig Covault, "Rocket Propulsion Tests Aimed at X-33, X-34 and Delta 4," *Aviation Week and Space Technology*, May 4, 1998, pp. 51–53; Ira F. Kuhn Jr., "Potential for Long Standoff, Low Cost, Precision Attack," in Defense Science Board 1996 Summer Study Task Force, *Tactics and Technology for 21st Century Military Superiority*, vol. 3 (Department of Defense, 1996), p. III-28; O'Hanlon, *Techno-

logical Change and the Future of Warfare, pp. 76–79; and Watts, *The Military Uses of Space,* p. 58.

50. Andy Pasztor and Anne Marie Squeo, "New Boeing Rocket Blasts Off in Long-Awaited First Launch," *Wall Street Journal,* November 21, 2002.

51. Craig Covault, "Delta IV Thrusts Boeing against Atlas V, Ariane," *Aviation Week and Space Technology,* November 25, 2002, p. 22.

52. See Robert Wall, "Costs Skyrocket," *Aviation Week and Space Technology,* November 24, 2003, p. 24.

53. Stew Magnuson, "Air Force Explores Balloon-Assisted Launches," *Space News,* January 13, 2003, p. 20.

54. Raytheon Corporation, "Ground-Based Midcourse Defense Exoatmospheric Kill Vehicle," 2002, available at www.raytheon.com/ekv.

55. American Physical Society, *Report of the APS Study Group on Boost-Phase Intercept Systems for National Missile Defense,* vol. 1, *Executive Summary and Conclusions* (July 2003).

56. Michael Sirak, "USA Works on Kinetic Energy Interceptor," *Jane's Defence Weekly,* January 8, 2003, p. 2.

57. See Daniel Kleppner, Frederick K. Lamb, and David E. Mosher, "Boost-Phase Defense Is a Challenge," *Space News,* September 1, 2003, based on the American Physical Society's 2003 study on boost-phase defense and American Physical Society, *Report of the APS Study Group on Boost-Phase Intercept Systems for National Missile Defense,* pp. 103–30.

58. Bob Preston and others, *Space Weapons, Earth Wars* (Santa Monica, Calif.: RAND, 2002), pp. 40–49.

59. Andrew Koch, "USAF Takes New Look at 'Big BLU'-Style Bomb," *Jane's Defence Weekly,* October 30, 2002.

60. "Incoming," *New York Times,* October 13, 2002, p. 3.

61. Simon P. Worden and Martin E. B. France, "Towards an Evolving Deterrence Strategy: Space and Information Dominance," *Comparative Strategy,* vol. 20 (2001), pp. 453–66.

62. Watts, *The Military Uses of Space,* pp. 53–54.

63. See "Outlook/Specifications: Spacecraft," in Aviation Week and Space Technology, *2002 Aerospace Source Book* (January 14, 2002), pp. 164–66.

64. Rumsfeld, "Report of the Commission to Assess United States National Security Space Management and Organization," pp. 20–21.

65. Worden and France, "Towards an Evolving Deterrence Strategy: Space and Information Dominance," p. 464.

66. Craig Covault, "USAF Technology Satellite Plays Tag with GPS Delta," *Aviation Week and Space Technology*, February 3, 2003, p. 39.

67. See Rumsfeld, "Report of the Commission to Assess United States National Security Space Management and Organization."

68. Personal communication from General Mike Hamel, 14th Air Force, Vandenberg Air Force Base, California, November 8, 2002.

69. Brian Berger, "Cost Projections for NASA's Next Space Telescope Exceed Budget," *Space News*, March 10, 2003, p. 1.

70. See Ivan Bekey, *Advanced Space System Concepts and Enabling Technologies for the 2000–2030 Time Period* (Annandale, Va.: Bekey Designs, July 7, 1998), pp. 13 and 24; Watts, *The Military Uses of Space*, p. 55.

71. Jeremy Singer, "U.S. Air Force Halts TechSat 21 Flight Demonstration," *Space News*, May 12, 2003, p. 9.

72. Watts, *The Military Uses of Space*, p. 81; Robert Wall, "Space-Based Radar Development Begins," *Aviation Week and Space Technology*, March 1, 1999, p. 33; O'Hanlon, *Technological Change*, pp. 39–40.

73. See Anne Marie Squeo, "Pentagon Finds Satellite Programs Seriously Flawed," *Wall Street Journal*, September 5, 2003.

74. Jeffrey T. Richelson, "The Satellite Gap," *Bulletin of the Atomic Scientists* (January–February 2003), pp. 49–54.

Notes to Chapter Four

1. Barbara Opall-Rome, "Iran Reportedly Plans to Launch Reconnaissance Satellite," *Space News*, October 14, 2002, p. 10.

2. See Paul D. Neilsen, "Antisatellite Weapons: A Strategic Analysis," National Defense Library Special Collections Paper (Washington: National Defense University, 1989), p. 12.

3. John Tirman, ed., *The Fallacy of Star Wars* (Random House, 1984), p. 194.

4. International Institute for Strategic Studies, *The Military Balance 2002/2003* (Oxford University Press, 2002). U.S. government estimates of China's military spending are typically about twice as great as

those of the International Institute for Strategic Studies. See Arms Control and Disarmament Agency, *World Military Expenditures and Arms Transfers 1996* (Government Printing Office, 1997), p. 65; for an explanation of the methodologies involved, see pp. 186–92, as well as International Institute for Strategic Studies, *The Military Balance 1995/96* (Oxford University Press, 1995), pp. 270–75.

5. William S. Cohen, "The Security Situation in the Taiwan Strait," Report to Congress pursuant to the FY99 Appropriations Bill (Department of Defense, 1999), pp. 9, 11; International Institute for Strategic Studies, *The Military Balance 1999/2000* (Oxford University Press, 1999), pp. 187–88; William S. Cohen, "Future Military Capabilities and Strategy of the People's Republic of China," Report to Congress pursuant to the FY98 National Defense Authorization Act (Department of Defense, 1998), pp. 15–16; and Edward B. Atkeson, "The People's Republic of China in Transition: An Assessment of the People's Liberation Army," Land Warfare Paper 29 (Alexandria, Va.: Institute of Land Warfare, Association of the U.S. Army, 1998), p. 11.

6. Gao Heng, "Future Military Trends," in Michael Pillsbury, ed., *Chinese Views of Future Warfare* (Washington: National Defense University Press, 1997), pp. 85–94; Dennis J. Blasko, Philip T. Klapakis, and John F. Corbett Jr., "Training Tomorrow's PLA: A Mixed Bag of Tricks," in David Shambaugh and Richard H. Yang, eds., *China's Military in Transition* (Oxford: Clarendon Press, 1997), pp. 225–60; Cohen, "The Security Situation in the Taiwan Strait," pp. 6, 11, 13; Cohen, "Future Military Capabilities and Strategy of the People's Republic of China," p. 8; Richard A. Bitzinger and Bates Gill, *Gearing Up for High-Tech Warfare?* (Washington: Center for Strategic and Budgetary Assessments, 1996); and Andrew N. D. Yang and Milton Wen-Chung Liao, "PLA Rapid Reaction Forces: Concept, Training, and Preliminary Assessment," in James C. Mulvenon and Richard H. Yang, eds., *The People's Liberation Army in the Information Age* (Santa Monica, Calif.: RAND, 1999), pp. 56–57.

7. Richard A. Bitzinger, "Going Places or Running in Place? China's Efforts to Leverage Advanced Technologies for Military Use," in Susan M. Puska, ed., *People's Liberation Army after Next* (Carlisle, Pa.: U.S. Army War College, 2000), pp. 9–54; Tim Huxley and Susan Willett, *Arming East Asia*, Adelphi Paper 329 (Oxford University Press, 1999), pp. 75–77; Lieutenant General Patrick M. Hughes, "Global Threats and Challenges: The Decades Ahead," Statement for

the *Congressional Record* (Washington: Defense Intelligence Agency, February 1999), p. 10; Avery Goldstein, "Great Expectations: Interpreting China's Arrival," *International Security*, vol. 22, no. 3 (Winter 1997–98), p. 46; John Wilson Lewis and Xue Litai, "China's Search for a Modern Air Force," *International Security*, vol. 24, no.1 (Summer 1999), p. 87; Eric McVadon, "PRC Exercises, Doctrine and Tactics toward Taiwan: The Naval Dimension," in James R. Lilley and Chuck Downs, eds., *Crisis in the Taiwan Strait* (Washington: National Defense University, 1997), p. 261; Kenneth W. Allen, "PLAAF Modernization: An Assessment," in Lilley and Downs, *Crisis in the Taiwan Strait*, pp. 232–40; Jonathan Brodie, "China Moves to Buy More Russian Aircraft, Warships, and Submarines," *Jane's Defence Weekly*, December 22, 1999, p. 15; and Office of Naval Intelligence, *Worldwide Challenges to Naval Strike Warfare* (Washington: U.S. Navy, 1996), p. 29.

8. Michael E. O'Hanlon, *Defense Policy Choices for the Bush Administration* (Brookings, 2002), pp. 154–203.

9. Thomas J. Christensen, "Posing Problems without Catching Up: China's Rise and Challenges for U.S. Security Policy," *International Security*, vol. 25, no. 4 (Spring 2001), pp. 5–40; and O'Hanlon, *Defense Policy Choices for the Bush Administration*, pp. 154–203.

10. See, for example, Andrew F. Krepinevich Jr., *A New Navy for a New Era* (Washington: Center for Strategic and Budgetary Assessments, 1996); and Andrew F. Krepinevich Jr., *The Conflict Environment of 2016: A Scenario-Based Approach* (Washington: Center for Strategic and Budgetary Assessments, 1996).

11. Bill Gertz, "Chinese Missile Has Twice the Range U.S. Anticipated," *Washington Times*, November 20, 2002, p. 3.

12. Department of Defense, "Annual Report on the Military Power of the People's Republic of China" (2002 and 2003), available at www.defenselink.mil.

13. Bruce G. Blair, *Strategic Command and Control: Redefining the Nuclear Threat* (Brookings, 1985), pp. 201–07.

14. See Tom Stefanick, *Strategic Antisubmarine Warfare and Naval Strategy* (Lexington, Mass.: Lexington Books, 1987), pp. 241–64; Robert E. Harkavy, *Bases Abroad: The Global Foreign Military Presence* (Oxford University Press, 1989), pp. 192–96; and Jeffrey T. Richelson, *The U.S. Intelligence Community* (Cambridge, Mass.: Ballinger, 1989), pp. 206–09.

15. Stefanick, *Strategic Antisubmarine Warfare*, pp. 38–41, 254–55.

16. Eric J. Labs, *Increasing the Mission Capability of the Attack Submarine Force* (Congressional Budget Office, 2002), pp. 5–7.

Notes to Chapter Five

1. Alexander G. Higgins, "China, Russia Ease Up on Space Arms," *Moscow Times*, August 8, 2003, p. 4.

2. See Rebecca Johnson, *Missile Defence and the Weaponisation of Space*, ISIS Policy Paper on Ballistic Missile Defense 11 (London: International Security Information Service, January 2003), available at www.isisuk.demon.co.uk; Jonathan Dean, "Defenses in Space: Treaty Issues," in James Clay Moltz, ed., *Future Security in Space: Commercial, Military, and Arms Control Trade-Offs*, Occasional Paper 10 (Monterey, Calif.: Monterey Institute of International Studies, 2002), p. 4; George Bunn and John B. Rhinelander, "Outer Space Treaty May Ban Strike Weapons," *Arms Control Today*, vol. 32, no. 5 (June 2002), p. 24; and Bruce M. Deblois, "Space Sanctuary: A Viable National Strategy," *Aerospace Power Journal* (Winter 1998), p. 41.

3. For a proposal along these lines, see Michael Krepon with Christopher Clary, *Space Assurance or Space Dominance? The Case Against Weaponizing Space* (Washington: Henry L. Stimson Center, 2003), pp. 109–10.

4. For an earlier, highly sophisticated argument along these lines, see John Tirman, ed., *The Fallacy of Star Wars* (Vintage Books, 1984).

5. James Clay Moltz, "Breaking the Deadlock on Space Arms Control," *Arms Control Today*, vol. 32, no. 3 (April 2002).

6. For a good discussion, see Krepon and Clary, *Space Assurance or Space Dominance?* pp. 114–24.

7. Krepon and Clary, *Space Assurance or Space Dominance?* p. 93.

8. For a summary, see David Mosher and Michael O'Hanlon, *The START Treaty and Beyond* (Congressional Budget Office, October 1991), pp. 34–35; Ivo H. Daalder, *Cooperative Arms Control: A New Agenda for the Post–Cold War Era*, CISSM Papers 1 (University of Maryland at College Park, October 1992), pp. 23–27.

9. Peter L. Hays, "Military Space Cooperation: Opportunities and Challenges," in Moltz, *Future Security in Space*, p. 42.

10. James Clay Moltz, "Reining in the Space Cowboys," in Moltz, *Future Security in Space,* pp. 62–64.

Notes to Chapter Six

1. Vice Admiral Lowell E. Jacoby, Director, Defense Intelligence Agency, Testimony, in "Current and Projected National Security Threats to the United States, Hearings before the Select Committee on Intelligence of the United States Senate, One Hundred and Eighth Congress, February 11, 2003" (Government Printing Office, 2003).

2. Jacoby, Testimony, p. 17.

3. See James Clay Moltz, "Reining in the Space Cowboys," *Bulletin of the Atomic Scientists* (January/February 2003), p. 66.

4. U.S. Space Command (before merging with Strategic Command), *Long-Range Plan* (1998), p. 21, available at www.spacecom.mil/LRP; Robert Wall and David A. Fulghum, "Satellite Self-Protection Gains Added Attention," *Aviation Week and Space Technology*, October 28, 2002, p. 68.

5. Wall and Fulghum, "Satellite Self-Protection Gains Added Attention," p. 68.

6. William B. Scott, "Innovation Is Currency of USAF Space Battlelab," *Aviation Week and Space Technology*, April 3, 2000, p. 53.

7. Andrew Koch, "U.S. Seeks Solutions to Space Threats," *Jane's Defence Weekly*, August 13, 2003.

8. Communication to author at Vandenberg Air Force Base, November 8, 2002.

9. Michael Sirak, "US Air Force Targets UAV Bandwidth Problem," *Jane's Defence Weekly*, July 31, 2002, p. 28.

10. David A. Fulghum, "It Takes a Network to Beat a Network," *Aviation Week and Space Technology*, November 11, 2002, p. 31.

11. Paul Rehmus, *The Army's Bandwidth Bottleneck* (Congressional Budget Office, August 2003), pp. ix–xiii.

12. Jeremy Singer, "U.S. Air Force Scales Back GPS Upgrade Plans," *Space News*, January 27, 2003, p. 8.

13. Phillip J. Baines, "Prospects for 'Non-Offensive' Defenses in Space," in James Clay Moltz, ed., *New Challenges in Missile Proliferation, Missile Defense, and Space Security*, Occasional Paper 12 (Monterey, Calif.: Monterey Institute of International Studies, July 2003), pp. 40–41.

14. It may be possible, at least in theory, to clean up electrons pumped into Van Allen belts after a nuclear explosion. In other words, it may be possible to reverse the so-called Christofilos Effect, specifically through the use of low-freqency kilohertz waves emitted from ground stations to make electrons "rain out" of orbit. This may help make low-altitude space usable within months instead of years—provided, of course, that subsequent nuclear explosions can be prevented and that new satellites can be orbited reasonably quickly to replace those that have been lost. See Ian Steer and Melanie Bright, "Blind, Deaf, and Dumb," *Jane's Defence Weekly*, October 23, 2002, pp. 21–23.

15. Ira W. Merritt, U.S. Army Space and Missile Defense Command, "Radio Frequency Weapons and Proliferation: Potential Impact on the Economy," Statement before the Joint Economic Committee, 105 Cong. 2 sess. (February 25, 1998); David A. Fulghum, "Microwave Weapons Await a Future War," *Aviation Week and Space Technology*, June 7, 1999, pp. 30–31; Carlo Kopp, "The E-Bomb—A Weapon of Electrical Mass Destruction" (Monash University, Australia, 1998); and Barbara Starr, "Russian Bomb-Disarming Device Triggers Concerns," *Jane's Defence Weekly*, March 18, 1998, p. 4.

16. Robert Wall, "Chinese Advance in Electronic Attack," *Aviation Week and Space Technology*, October 28, 2002, p. 70.

17. See Tom Wilson, "Threats to United States Space Capabilities," paper prepared for the Commission to Assess United States National Security Space Management and Organization (Washington, 2001), pp. 41–46.

18. U.S. Space Command, *Long-Range Plan*, p. 24.

19. I am indebted to Richard Garwin for this observation (personal communication, August 25, 2003).

20. John A. Tirpak, "Challenges Ahead for Military Space," *Air Force Magazine* (January 2003), pp. 25–26.

21. Marcia S. Smith, *U.S. Space Programs: Civilian, Military, and Commercial* (Congressional Research Service, September 16, 2003), p. 13; Office of the Under Secretary of Defense (Comptroller), *RDT&E Programs (R-1)*, fiscal years 2003 and 2004 respectively, available at www.defenselink.mil.

22. Benjamin S. Lambeth, *Mastering the Ultimate High Ground* (Santa Monica, Calif.: RAND, 2003), p. 88.

23. Jeremy Singer, "U.S. Developing Ground-Based Systems to Counter Satellites," *Space News*, June 30, 2003, p. 6; Jeffrey Lewis,

"President Bush's 2004 Budget Request: Implications for Space Weaponization," draft, Center for International and Security Studies, University of Maryland, June 1, 2003 (cited with the permission of the author); and "Air Force Budget Request for 2004" (specifically, program element PE 0603401F), available at www.vs.afrl.af.mil/Factsheets/XSS10.html.

24. Strategic Command recognizes as much. See U.S. Space Command, *Long-Range Plan*, p. 44.

25. Michael Sirak, "Pentagon Eyes Near-Term Ability to Block Satcom," *Jane's Defence Weekly*, July 24, 2002, p. 8.

26. U.S. Space Command, *Long-Range Plan*, pp. 46, 63.

INDEX

ABL. *See* Airborne laser

ABM Treaty. *See* Anti-Ballistic Missile Treaty

Afghanistan war, 3, 4, 38

AFSATCOM satellites, 46

Airborne laser (ABL), 27, 72–76, 110, 113, 136, 137, 139

Aircraft, 26, 129, 137

Air Force, U.S., 50, 77, 136

Air Force Research Laboratory, 125

Anti-Ballistic Missile (ABM) Treaty, 2–3, 15, 22, 23, 112

Antisatellite (ASAT) programs, 8–9; availability, 120; bans and prohibitions, 12, 23, 107–13; capabilities, 8, 18, 26–28, 129, 134; debris-producing, 106, 112–13, 116, 128, 132, 138; hardening and defense against, 25, 46, 68–70, 88, 93, 122–30; hit-to-kill technologies, 65, 83–84; hunter-killer satellites, 128; intercept schemes, 66; keep-out zones, 114, 124; kinetic energy, 13, 28, 85; kinetic kill, 53; microsatellite, 27–28, 86–88, 129, 130; microwave weapons, 69–70, 87, 124, 127; missile defense systems and, 136–38; nuclear detonations, 67–70; research and development, 133–42; satellite vulnerabilities, 99–100; space mines, 87, 128; space weaponization and, 18, 21, 107, 122–42; testing, 138–39; verification, 107–08, 110, 111, 112, 113, 121. *See also* Lasers; *individual systems and countries*

Arab Satellite Communications Organization, 59

165

Argentina, 57, 98

Ariane rocket, 57

Arms control: accords and treaties, 2–3, 105–08, 120–21, 141; bans and prohibitions, 109–13, 121, 130, 138; confidence-building measures, 113–15; deterrence, 130–31; informal unilateral restraints, 115–17; recommendations, 138–39; verification, 107–08, 110, 111, 112, 113, 121, 130

Army, U.S., 136

ASAT programs. *See* Antisatellite programs

Asia Satellite Telecom Company, 59

Athena rocket, 47

Atlas rocket, 47, 83

Ballistic missile defense: brilliant pebbles concept, 84–85, 120; in cold war, 9; deployment, 9; hit-to-kill technologies, 65, 83–84; midcourse intercept system, 27, 136, 137, 138; space-based, 28, 53, 110; U.S. systems, 119–20. *See also* Lasers; Missiles; *individual systems*

Ballistic missiles, 5, 16–17, 27, 67, 74, 110, 113, 119, 137; ICBMs, 27, 68, 74; SLBMs, 27. *See also* Ballistic missile defense; Missiles

Boeing, 83

Brazil, 56, 57, 86, 99

Bush, George H. W., 13

Bush, George W.: ABM Treaty and, 112, 121; laser research under, 13; Putin and, 22; space policy under, 13–14, 24

Bush (George H. W.) administration, 116

Bush (George W.) administration, 15, 24, 46–47

Canada, 17, 58, 86

CFE Treaty. *See* Conventional Armed Forces in Europe Treaty

Challenger space shuttle, 42

Chemical lasers, 71–72, 73, 77, 79

China. *See* People's Republic of China

CIS. *See* Commonwealth of Independent States

Clementine II program, 13

Clinton, Bill, 12–13, 14, 38, 112

Clinton administration, 73, 84, 115, 136, 137

Cold war: antisatellite weapons, 9; arms control, 113–14; arms race, 133; development of mobile monitoring terminals, 47; nuclear weapons, 116; end of, 12; use of satellites, 2–3; U.S. missile defenses, 133

Columbia space shuttle, 47

Commercial and civilian satellites, 40–41, 54, 56, 122, 127

Commercial Remote Sensing Policy, 38

Commission on Outer Space, 21, 120, 122

Commonwealth of Independent States (CIS), 36. *See also* Russia; Soviet Union

Communications: satellites, 3–4, 43, 46, 51, 56, 63, 100, 120, 122, 125–26; systems, 80–82, 114

Computers and software, 27, 75

Conference on Disarmament, 17, 106, 109

Conventional Armed Forces in Europe (CFE) Treaty, 3, 12

Cosmos rocket, 53–54

Cox report (*1999*), 76

Cuba, 2, 125

Daschle, Tom, 16

Dean, Jonathan, 12, 109

Defense, Department of, 78, 125, 135

Defense Intelligence Agency, 95, 120

Defense Meteorological Satellite Program, 43

Defense Satellite Communications System (DSCS), 46, 51, 126

Defense Science Board, 78

Defense Support Program, 46, 124

Delta rocket, 47, 83

Democratic position on space, 16

Discoverer II program, 53, 89

DSCS. *See* Defense Satellite Communications System

Early and attack warning satellites, 43, 46, 54, 124–25

Earth: atmospheric effects, 73, 75, 76, 77–78; curvature, 77; gravitational pull, 3031, 33; size, 30. *See also* Geosynchronous orbit; Low-Earth orbit; Medium-Earth orbit

Earth Watch, 58

EC-*135*. *See* Aircraft

Economic issues, 16, 35, 133–38

EELV. *See* Evolved expendable launch vehicle

Electromagnetic pulses, 68

Eros satellite, 59

European Space Agency, 56, 57

European Telecom Satellite Organization, 59

Evolved expendable launch vehicle (EELV), 47, 83

F–*15*s. *See* Aircraft

Falklands War, 93, 98

FIA. *See* Future Imagery Architecture

Fiber-optic cable, 16, 35, 81, 129

FLSAT, 46

France, 18, 56, 57, 86, 94, 99

Frequencies, 51

Future Imagery Architecture (FIA), 51, 63, 89

Garwin, Richard, 123

Galileo navigation system, 57

Geosynchronous orbit: launching of satellites, 50; region and range, 30–31, 32, 68; rockets for, 33; satellites in, 35, 38, 42,

43, 46, 51, 54, 56; timing and speed, 32, 33; Van Allen radiation belts and, 34; vulnerabilities of GEO satellites, 67–68

Germany, 47, 86, 99

Global broadcast systems, 46

Global positioning system (GPS) satellites, 3, 5, 26, 38, 42, 46, 51, 57, 63, 126

Globalstar, 59

GLONASS navigation system, 54

GPS. See Global positioning system satellites

GSLV rocket, 58

Hardening of satellites, 25, 46, 68–70, 88, 93, 122–30, 141

Helios satellite, 56

High-energy lasers, 65, 71–77, 123, 130, 135

H-2 and H-2A rockets, 57

ICBMs. See Intercontinental ballistic missiles

Ikonos satellite, 58–59

Imaging, 12, 34, 38, 43, 46, 51, 54, 56, 57, 58–59, 120, 126

Incidents at Sea agreement, 114

India, 58, 76

Indonesia, 58

INF Treaty. See Intermediate-Range Nuclear Forces Treaty

Intercontinental ballistic missiles (ICBMs). See Ballistic missiles

Intermediate-Range Nuclear Forces (INF) Treaty, 12

International Maritime Organization, 59

International Telecom Satellite Organization, 59

Iran, 13, 91

Iraq. See Operation Desert Storm; Operation Iraqi Freedom

Israel, 57, 58, 59, 86

Japan, 47, 50, 57, 94

JDAM. See Joint direct attack munition

JSTARS. See Joint Surveillance Target Attack Radar System

Johnson Island, 69

Joint direct attack munition (JDAM), 3, 4

Joint Surveillance Target Attack Radar System (JSTARS), 26

Jumper, John, 8

Jumpseat satellites, 46

Keyhole (KH) satellite systems, 33–34, 43, 46

Kosovo war, 3, 4

Lacrosse satellite, 33–34, 43, 46

LANDSAT, 58

Lasers: all-gas-phase iodine laser, 80; antimissile and -satellite capabilities, 73–77, 122; attacks by, 25, 123; beam strength and range, 74, 77–78, 109; communications systems, 81–82; construction requirements, 79–80; damaging doses, 71–72; dazzlers, 13, 136; Earth's atmosphere and, 73, 75, 76, 77–78, 79, 81; free-electron lasers, 72, 80; laser satellite constellations, 124; low-energy lasers, 76; mirrors and optical systems, 78–79; nuclear-pumped lasers, 76; pulsed lasers, 71; satellite communications systems, 25, 125; solid-state lasers, 72; wavelengths, 77–78. *See also* Airborne lasers; Chemical lasers; High-energy lasers; Space-based lasers; Mid-infrared advanced chemical laser

LEO. *See* Low-Earth orbit

Livermore labs, 82

LK-*1* rocket, 58

Lockheed Martin, 83

Long March rocket series, 56

Low-Earth orbit (LEO): ASAT debris, 112; protection and hardening of satellites in, 123, 127; region and range, 30, 68; rockets for, 33; satellites in, 8–9, 12, 26, 27, 35, 38, 42, 43, 51, 53, 54, 56, 127; timing and speed, 33; Van Allen radiation belts and, 34, 68–69; vulnerabilities of satellites in, 67, 68, 136, 137

Magnum satellite, 46

Medium-Earth orbit (MEO): hardening of satellites in, 127; region and range, 30; satellites in, 38, 42, 46; Van Allen radiation belts and, 34, 127

Microsatellites, 65, 85–89, 100, 109, 121, 125, 130; ASAT programs, 27–28, 86–88, 129, 130; Chinese, 86, 87, 92; Russian, 86, 87; U.S., 86, 87, 88

Microwaves, 69–70, 87, 124, 127, 140

Mid-infrared advanced chemical laser (MIRACL), 13, 27, 73, 77

Military issues: air-breathing capabilities, 26; future conflicts, 91–104; militarization of space, 1–2; military organization, 50; negation, 14–15; space-to-Earth weapons, 28

Military Strategic and Tactical Relay (MILSTAR), 46, 51, 123, 126

Military use of satellites, 1–2, 3–4, 21, 22–23, 38–39, 42, 43–46, 54, 56, 63, 88, 125–26; commercial satellites and, 46–47, 125; DSCS, 46; GPS, 3

MILSTAR. *See* Military Strategic and Tactical Relay

MIRACL. *See* Mid-infrared advanced chemical laser

Missile Defense Agency, 74, 77, 84

Missiles: antisatellite capabilities, 27, 110, 119, 121; Chinese, 96, 98; defenses, 28, 84–85, 136–37; early warning satellites, 43, 46; hypersonic missiles, 85; interceptor missiles, 9, 84–85; nuclear-armed or -tipped, 50, 67, 92, 122, 130; recon-naissance-strike complex, 101;

SCUDs, 68, 71, 72, 73; Sidewinders, 73; tracking, 51, 65. *See also* Anti-Ballistic Missile Treaty; Ballistic missile defense; Ballistic missiles; Rockets and rocketry

Mobile User Objective System (MUOS), 51

Molniya rocket, 53–54

MUOS. *See* Mobile User Objective System

National Command Authorities, 14

National Imagery and Mapping Agency, 38

National Reconnaissance Office, 124

National Reconnaissance Organization, 15

Navy, U.S., 75

Nuclear weapons, 8, 67–70, 116, 126–27

Observation satellites, 54, 56

Onyx satellite systems, 43

Open Skies Treaty, 3

Operation Desert Storm, 3, 4, 58, 140

Operation Iraqi Freedom, 3, 4, 5, 46

Orbcomm, 59

Orbimage, 58–59

Orbital Sciences, 58

Orbits: altitude of, 30; elongated, 46; objects in Earth orbit, 36–37; of satellites, 30–34, 35,

38, 42–43; shape of, 32, 38, 43. *See also* Geosynchronous orbit; Low-Earth orbit; Medium-Earth orbit

OrbView-*3* and -*4* satellites, 59

Outer Space Treaty, 20, 107, 109

P-*3*. *See* Aircraft

Pakistan, 86

Peaceful Nuclear Explosions Treaty, 3

Pegasus rocket, 47

People's Republic of China (PRC): future Taiwan Strait conflict, 91–104, 110–11, 131; manned space program, 54; micro-satellites, 86, 87, 92; militari-zation of space and, 17, 99; military capabilities, 93–104; monitoring of, 2; People's Liberation Army, 54, 95; satellite launch services, 33, 54; satellites, 39, 54, 56, 58, 61; space capabilities, 5, 8, 13, 18–19, 92–93, 98–104, 127; treaty negotiations, 106; U.S.-China relations, 22, 24, 92; use of lasers, 76–77

Picosatellites, 86

Political issues, 16, 21–22, 28, 103, 104, 106, 134–35

Portugal, 86

PRC. *See* People's Republic of China

Presidential decision directive 23, 38

Putin, Vladimir, 22

Quickbird satellite, 58

Radar, 50, 51, 53, 56
Radiation. *See* Lasers; Microwaves; Van Allen radiation belts; X-rays
Radio energy, 127
Research and development, 133–38
Rockets and rocketry: development of, 63, 90; EELV, 47, 83; space launching, 33–34, 48–49, 82–84; space shuttle, 47. *See also* Missiles; *individual systems*
Rumsfeld, Donald, 13–14, 15, 16
Russia: Chinese relations, 56, 76, 99; manned space program, 54; microsatellites, 86, 87; military spending, 5, 94; notification of space launches, 12; satellites, 35, 38, 53–54, 58, 61; space budget, 53; space facilities, 53; treaty negotiations, 106; use of lasers, 76; U.S. relations, 22, 24, 115; weaponization of space and, 17. *See also* Commonwealth of Independent States; Soviet Union

SALT II. *See* Strategic Arms Limitation Talks
Sanctuary, space as, 8–9, 12, 15, 18, 22
Satellites: backups and alternatives, 128–30, 137; communications capabilities, 35, 51; costs, 33, 89; current

military and commercial, 34–42, 69, 125, 127; data transmission, 4, 56, 125–26, 149n14; defensive or offensive use, 5, 18, 93; de-orbit of, 117; effects of Earth's curvature, 46; ground control stations, 47; hardening, 25, 46, 68–70, 88, 93, 122–30, 141; jamming, 125–26, 136, 137, 140; launching, 32–35, 47–49, 55, 115; passive and self-defense measures, 25–26; satellite swarms, 88–89; speeds, trajectories, and revolutions, 30–33; Van Allen radiation belts and, 34; vulnerabilities, 99–101; weight and size, 33–35. *See also* Microsatellites; Picosatellites; *individual systems*
SBIRS. *See* Space-based infrared satellite systems
SBLs. *See* Space-based lasers
Schriever Air Force Base, 47, 50, 125
Sea Launch, 57
September *11, 2001,* 22, 24, 35
Shavit rocket, 58
Shtil rocket, 53–54
Signals-intelligence satellites, 46, 51
SLBMs. *See* Submarine-launched ballistic missiles
Smith, Robert, 13
SOSUS array, 101, 102
Soviet Union: antisatellite capability and tests, 9, 10–11, 130; dissolution of, 53; listening

to, 46; monitoring of, 2; nuclear weapons, 116; space program, 2, 92–93

Soyuz rocket, 53–54

Space and Missile Defense Command, 50

Space-based infrared satellite (SBIRS) systems, 51, 63, 89

Spaced-based lasers (SBLs), 53, 77–80

Space Command, 14–15, 47, 50

Space Imaging Corporation, 38, 58

Space shuttle, 42, 47

Space Tracking and Surveillance System (STSS), 51

Space Wings, 50

SPOT, 56, 58

Sputnik, 2

Start rocket, 53–54

START I and II. *See* Strategic Arms Reduction Treaties

Strategic Arms Limitation Talks (SALT) II, 3

Strategic Arms Reduction Treaties: START I, 3; START II, 12

Strategic Command, 15, 129

STSS. *See* Space Tracking and Surveillance System

Submarine-launched ballistic missiles (SLBMs). *See* Ballistic missiles

Surrey Satellite Technology, 86

Taiwan, 76, 91–104, 110–11

Technology advances, 62–63, 89

TechSat21, 89

Teets, Peter B., 15

Telephone services. *See* Fiber-optic cable

Telescopy, 50

Terrorism. *See* September *11, 2001*

Theater missile defense (TMD), 12, 23

Thermal blooming, 76

Threshold Test Ban Treaty, 3

Titan rocket, 34, 47

TMD. *See* Theater missile defense

Treaties: bans and prohibitions, 105–13; confidence-building measures, 113–15; United Nations and, 105–06. *See also individual treaties*

UAVs. *See* Unmanned aerial vehicles

Ukraine, 33, 58, 99

United Kingdom, 47, 86, 87, 94, 99

United Nations (UN): Conference on Disarmament, 17, 106, 109; reconnaissance satellites, 2; treaties, 105–06; weaponization of space, 17, 22, 121

United States: antisatellite capability, 9, 12–13, 27, 63, 65, 105–17, 121–22, 130–42; arms control measures, 23–24, 105–17; dominance in space, 29, 61–62, 119–42; future satellite capabilities, 51, 53, 103–04; international views of, 22; launch facilities, 47–49; microsatellites, 86, 87, 88;

military and space spending, 5,
6–7, 133–38, 155n38; missile
defense system, 110; notification
of space launches, 12; and
regime change, 67; Russian
relations, 22, 24, 115; satellites,
35, 38, 42–53, 120, 122–30;
space policy and program, 2–3,
16, 119, 120–22, 141; space
surveillance, 50, 75, 124–25;
space system survivability
measures, 64; threats to, 13–14;
UN and, 105–06; weaponi-
zation of space, 16–24, 120–22,
132–42; weapons of mass
destruction, 132–33
Unmanned aerial vehicles (UAVs),
3, 26, 82, 129

Van Allen radiation belts, 8, 34,
68–69, 127, 163n14
Vega rocket, 57
VLS-1 rocket, 58

Watts, Barry, 18
Weapons of mass destruction
(WMD), 132–33
Weather satellites, 38, 43
White Cloud ocean reconnaissance
satellite, 43, 46
Wideband Gapfiller satellites, 51
Worden, Pete, 15
World War II, 116

X–rays, 68, 71

Zenit rocket, 57